Sustainable Urbanization: Challenges and Opportunities

Sustainable Urbanization: Challenges and Opportunities

Editor: Matthew Perez

MURPHY & MOORE
www.murphy-moorepublishing.com

Murphy & Moore Publishing,
1 Rockefeller Plaza,
New York City, NY 10020, USA

Visit us on the World Wide Web at:
www.murphy-moorepublishing.com

ISBN: 978-1-63987-514-6 (Hardback)

Cataloging-in-Publication Data

Sustainable urbanization : challenges and opportunities / edited by Matthew Perez.
p. cm.
Includes bibliographical references and index.
ISBN 978-1-63987-514-6
1. Urbanization--Environmental aspects. 2. Sustainable urban development.
3. Urban ecology (Sociology). 4. City planning--Environmental aspects.
I. Perez, Matthew.
HT241 .S87 2022
307.76--dc23

Table of Contents

Preface

Urbanization refers to the process by which towns and cities are formed and expand. The populations of cities are increasing faster than the ability of local and national governments to plan and manage these changes. This requires an urgent need of sustainable urbanization to properly manage the urban development. Well-managed cities have the potential to serve as engines of economic progress. Sustainable urbanization is also essential for promoting stability and preventing violence, crime and extremism. It also plays a significant role in strengthening rural-urban linkages and improving agricultural development. This book brings forth some of the most innovative concepts and elucidates the unexplored aspects of sustainable urbanization. It presents researches and studies performed by experts across the globe. This book, with its detailed analyses and data, will prove immensely beneficial to professionals and students involved in this area at various levels.

The information shared in this book is based on empirical researches made by veterans in this field of study. The elaborative information provided in this book will help the readers further their scope of knowledge leading to advancements in this field.

Finally, I would like to thank my fellow researchers who gave constructive feedback and my family members who supported me at every step of my research.

Editor

1

Brownfield Redevelopment in Turkey as a Tool for Sustainable Urbanization

Gökçen Kılınç Ürkmez

Abstract

Brownfield redevelopment is becoming a major planning issue with its environmental, social, economic, and spatial dimensions all around the world. As the attempts to manage the risks and costs associated with brownfields improve and the perception and awareness through the society increases, more stakeholders are put effort to achieve a broader range of environmental, social, and economic objectives under the concept of sustainable development. Since the mid-1980s, sustainable development has become widely discussed approach for integrating environmental needs with economic and social ones in human development. In this context, brownfields left over from industrialization provide an opportunity to address sustainable developments through urban planning approaches. As a result of the emergence of the sustainable development and brownfield regeneration agendas, there has been increased debate over the concept of "sustainable brownfield regeneration." Turkey has relatively a short history of industrialization; thus, brownfields are a new phenomenon which started to occur in the middle of 1980s. The brownfield policies in Turkey, generally discussed in context with urbanization and heritage preservation. However, their appearance and redevelopment have much related with the privatization and neoliberal policies which have had great effects in the country economic and social structure. Besides, EU candidacy have stimulated the compliance of Turkish legislation to the European standards since the beginning of last decade. Environmental and historical preservation and administrative decentralization constitute the significant subjects related to brownfields redevelopment. Cities in Turkey struggle with many economic, social, environmental, and politic problems. At this point, brownfields represent opportunities in order to obtain sustainability and increase the living standards especially in urban areas. This work aims to put forward the brownfield related policies in Turkey and to determine the main obstacles in brownfields redevelopment, the essential policies and strategies which can be transferred from Western countries experience and the essential steps which must be taken at the early stages of deindustrialization and decentralization for Turkey in context with sustainable urbanization.

Keywords: Deindustrialization, brownfields redevelopment, sustainable development, privatization, environmental policies

1. Introduction

In recent years, there has been a strong emphasis on brownfield redevelopment as a type of urban regeneration with a strong link to sustainable development in developed countries. As Dixon [1] indicates, such an approach highlights the importance of reusing and recycling brownfield sites in order to improve the urban environments and to release development pressures.

However, brownfields are generally not economically competitive when compared with greenfields without any public intervention. There are economic, social, and environmental barriers in returning the brownfields to beneficial uses. In spite of all problems, these sites may include, they are also regarded as an opportunity in providing sustainable urban development and to bring extra values into poor-quality lands and declined urban parts.

Sustainability has become a major issue in development strategies since the end of 1980s. As it is well known, in Brutland Report sustainable development is defined as "development that meets the needs of the present without compromising the ability of future generations to meet their own needs" [2]. At this point, it is necessary to think about whether all the brownfields regeneration projects really obtain sustainable development in the cities. What are the driving forces, policies, and handicaps in returning these areas into beneficial uses? Such a thought impulse us to ask what are the essential strategies in sustainable brownfield regeneration? Before determining the sustainable brownfield regeneration, it may be useful to look at the definitions about brownfields in all around the world.

2. Brownfield definitions

In the beginning of brownfield-related researches, one important problem is the lack of a common definition accepted by all countries in the world. The term "brownfield site" has different meanings in different countries. According to USEPA's renewed official definition brownfield site is "real property, the expansion, redevelopment, or reuse of which may be complicated by the presence or potential presence of a hazardous substance, pollutant, or contaminant".[1] Unlike the EPA's definition, brownfields are more referred to "previously developed lands" than "contaminated sites" in UK and are accepted as all abounded, idled,

[1] Small Business Liability Relief and Brownfields Revitalization Act, 2002.

or underused properties with fixed infrastructure and developed surface on site regardless of whether contamination is present or not [3]. In fact, most of the European countries do not determine the brownfields directly related with the contamination on the site [4]. But this does not mean that contamination is neglected in the brownfield redevelopment. Whatever the definition, brownfields can be characterized with some common features such as economic failure, difficulties in attracting new investors, high unemployment rate, negative effects on urban life, social conflicts, consumption of greenfields [5].

Unfortunately, lack of commitment on the term of brownfield makes difficult to determine the amount and dimension of the problem and also to designate common strategies for the sustainable development.

In spite of these complexities, the organizations such as RESCUE (Regeneration of European Sites in Cities and Urban Environments) try to define a common strategy and recommendations for EU countries through conducted analysis among the members. The main objective of RESCUE is to make brownfield redevelopment a key part of European Union's strategy on sustainable urban development and to show the EU what tools are necessary to maintain sustainability [6].

Another brownfield organization, funded by the Environment and Climate Programme of European Commission, CLARINET's (Contaminated Land Rehabilitation Network for Environmental Technologies) primary objectives are to develop technical recommendations for sound decision-making on the rehabilitation of contaminated sites in Europe and to identify research and development needs, in particular in relation to the EC Fifth Framework Programme [16]. CLARINET developed the concept of Risk Based Land Management as a step forward toward an integration of sustainable soil quality, the protection of water, and land-use management in environmental policy. In a report of CLARINET on brownfields, published in 2002, it is indicated that brownfields are spatial planning and ecological problem in relation to human health and it is proposed that successful brownfield redevelopment needs to have an integrated approach that addresses environmental and spatial planning problems simultaneously [6].

From this point of view, it must be noted that the main objective of brownfield redevelopment is not reuse or reintegration of land. It aims to supply a combination of social, environmental, and economic benefits through sustainable development perspective [7].

As a result of the emergence of the sustainable development and brownfield regeneration agendas, there has been increased debate over the concept of "sustainable brownfield regeneration". In this context, Dixon [1] tries to determine a number of sustainable brownfield development objectives to develop actions at the site level. These are minimizing the use of resources, minimizing pollution, protecting biodiversity and the natural environment, protecting the industrial heritage, and protecting the cultural environment.

It is possible to designate additional objectives related with the countries' and cities' visions and economic and environmental priorities such as, job creation, quality of life improvement, reducing urban sprawl.

3. Brownfield redevelopment in Turkey

The term "brownfield" or, as in the Turkish case, "former industrial area" has been recently introduced to Turkish agenda and is often related to the process of industrial decentralization and privatization of government-owned firms through state's liberalization policies. On the other hand, as the result of European Union membership attempts, many manufacturing industries had to stop their operations due to the impact of establishing the customs union between Turkey and the EU after 1995.

As a new phenomenon, brownfields redevelopment has been subject to academic research in Turkey since the beginning of the twenty-first century. The recent research about brownfields redevelopment in the country generally handled the issue from historical heritage preservation perspective with very limited and partial assessments.

Most of these scholarly studies deal with the preservation of industrial plants and buildings on the brownfields as important indicators of the former industrial identity of the cities and approach the issue as a design and conservation problem [8–22]. Almost all of these papers are concerned with brownfield sites at metropolitan areas such as Istanbul, Ankara, Bursa, Eskisehir, Izmit. As a general concluding remark, the political and legal uncertainties are emphasized in these writings. Koksal [19] also focuses on the insufficiency of brownfield-related inventory and highlights the necessary of analysis of the documentation for the old industrial plants as a form of historical archaeology. Asiliskender [9] argues in his Ph.D. dissertation the importance of the protection of these lands as symbols of Turkish modernity. Some of this research evaluates the brownfield redevelopment issue through urban regeneration concepts and policies and tries to put forward some proposals for alternative uses in order to connect these sites with other urban functions [23, 17, 24, 16, 18, 25, 15, 26, 27].

The other remarkable issue in these studies is the lack of interest in contamination- and environmental-related issues. Contamination is mentioned as an obstacle only in the paper of Turer Baskaya [28] and with a very limited focus. Another significant study which uses the term "brownfield" is the Ergen's Ph.D. Dissertation. Ergen [29] analyses the issue in terms of sustainable development with a focus on environmental impacts of abandoned lands from a perspective of landscape architecture in a shrinking city of Turkey.

There is no doubt that Turkish urban development and related issues are very different from Western countries. But in globalized world, there is a need to understand the specific and worldwide problems and to develop the most proper solutions to eliminate them.

Turkey, as a centralized country which is shifting its policies toward administrative decentralization and EU membership, has many difficulties in sustainable brownfield redevelopment, as in other regeneration projects. The complexity of brownfield redevelopment requires substantial efforts in the planning process to investigate contamination on the sites, to estimate market demands and to communicate with stakeholders. However, in Turkey, this complicated process has been challenged by rapid urbanization for decades, and the roots and policies for brownfields have had a process distinct from United States and other Western countries experiences [30]. Although there are many similarities in terms of globalization, technological changes and the liberal economic policies following after the 1980s, the historical perspective and driving forces which have caused the brownfields are different.

3.1. Policies, strategies, driving forces

Brownfield-related policies in Turkey represent the impacts of privatization and neoliberal policies after 1980s. In addition to the central role of the government, the strategies and priorities of local governments through their ideologies reshaped the cities, especially İstanbul. There is no doubt that the accomplishment of a sustainable brownfield redevelopment is directly related with the economic/social structure and environmental priorities of the countries. Turkey, as a developing country has been in a dilemma between economic development and environmental protection. The distinctions in industrialization process, social, economic, and politic structure necessitates to evaluate the sustainable brownfield redevelopment issue in its own dynamics.

3.1.1. Environmental policies

When compared with United States and Europe cases, the environmental dimension of brownfield redevelopment is much neglected in Turkey. Turkey started to discuss the environmental issues by the early years of 1970s. Over the last 45 years, it can be said that Turkey has made great progress in creating mechanisms to address its environmental problems: The 1982 Constitution recognizes the right of citizens to live in a healthy and balanced environment; an Environment Act was passed in 1983; the Ministry of Environment was created in 1991; public awareness and demand for a clean environment are growing; and active non-governmental environmental organizations are emerging [31]. Furthermore, in another Article of the constitution, environmental rights are not only defined as part of states purview, but also a duty for every citizen. Despite these positive developments, environmental issues have not been adequately incorporated into economic and social decisions [32].

The Environmental Law which came into force in 1983 endorses the "polluter-pays principle" which is the main principle of brownfield related legislation in USA and the environmental regulation in EU and handles environmental issues on a very broad scope [33]. There is not a specific regulation about brownfields even though the existence of many related laws and regulations which govern the environment and land use in Turkey.

In Turkish environmental legislation, Regulation for the Control of Soil Contamination and for the Particularly Contaminated Lands (RCSCPCL), taken effect in 2010, is very important in terms of brownfields issues. The aim of the regulation was to prevent the contamination of land, to find out the potential and probable contaminated lands and sectors, and to determine the clean-up and monitoring principles with respect to sustainable development purposes.

According to the RCSCPCL, the lands and plants which are contaminated as a result of disaster or accident, and which are determined as hazardous for the environment after the controls are recorded in the Hazardous Land List. The remediation of these sites requires the approval of "site analysis reports," "evaluation of the situation of land," and "risk reports and the clean-up activities reports" by the Province Directorate of Environment and Urbanization.

Moreover, in 1993, Environmental Impact Assessment (EIA) was enacted according to the United States and European procedures, and it was improved in 1999 according to European EIA directive. Even though the recent laws and regulations have raised the public awareness in environmental issues, there is still a big gap in the Turkish legislation in terms of liability and financial issues. However, it is evident that the further developments and regulations should be established the environmental roots of the country especially in terms of sustainable development.

3.1.2. EU candidacy

EU accession and candidacy can be considered as the most important driving force for the improvement of Turkish environment legislation and public awareness in sustainable brownfield redevelopments. Since the environment is one of the priority areas for Turkey's membership preparations, several new regulations put into force in the last two decades. Since EC-Turkey Association Council implementing the final phase of the Customs Union, there have been continuous efforts on the part of Turkey to harmonize its legislation with the EU *acquis communautaire* [34]. The candidacy indicates the priority areas for Turkey, and environment is one of them (in addition to other politic and administrative areas).

According to Budak [35], Turkey had at the turn of the century incorporated the environmental concerns and policies of the EU into its legal structure to a great extend as a non-member state. And, according to the findings of a research conducted in 2001, Turkey had made a great deal of progress in regard to the approximation of environmental *acquis* in some areas such as textile and chemicals [36].

However, compliance to EU rules sometimes can be costly for the national economy, and therefore, some challenges can be appeared in the society. EU environmental rules require the recognition of shared responsibilities. Actually, beyond the EU standards and profit maximization concerns, private actors must develop a new sense of responsibility for environment and sustainable urbanization [37].

EU memberships in Turkey still a question of socio-economic development and a slogan for modernization. In a broader sense, EU affects both legal and political structures and the public opinion. It seems that in short term, the impact of environmental policy on sustainable development-related issues such as environment, urbanization, growth, regeneration are likely to become more popular subjects in the public opinion and on the agenda of policy makers in Turkey. It can be said that, in spite of the improvements in environmental policy with the impulse of EU candidacy, further attempts are required in order to reach the developed counties standards.

3.1.3. Privatization policies

One of the main differences for the brownfields-related issues between Turkey and Western countries is coming from the liberal policies of 1980s which is very effective and cause huge changes in Turkish government system. Although the brownfields are appeared as the results of urban sprawl and deindustrialization in developed countries, in Turkey most of the

brownfields areas have become sites which need urgent solutions after the implementation of privatization policies.

As previously mentioned, privatization and governmental decentralization have been the main reasons for the appearance of brownfields in most of Turkish cities. Due to the fact that central policies have shifted to more decentralized structure in Turkey, many industrial plants which were established and operated by central government were privatized since the midst of 1980s.

In Turkey, Public Enterprises (PEs) had a significant role in the economic development of the country from the beginning of Republican Period. PEs emerged during the 1930s not for ideological but pragmatic reasons, mainly owing the lack of private capital accumulation and initiative to give the first impetus to development [38]. Most of them are large-scale investments and operations and used by the governments in order to create the necessary industrial infrastructure, to prevent unemployment and to reduce regional inequalities between the regions [38]. However, after starting the implementation of privatization policies, central government did not put any effort to revitalize or to enhance the technological capacity of these industrial enterprises. As a result, most of them closed and the sites remained idle.

In fact, today, the privatization practices are seen in many countries whose economies are called liberal, socialist, or mixed [38]. According to UNESCAP's (United Nations Economic and Social Commission for Asia and Pacific) [41] point of view *"the private sector is a key environmental actor, as its enterprises generally perform better than parastatals (government-owned companies), treating almost twice as much of their wastewater and spending less than half as much for final disposal."* Actually, in Turkish practice, it cannot be said that private enterprises are more environment-friendly than the government-owned ones.

The transfer of industrial areas by the impulse of privatization policies and globalization has caused vital changes especially in metropolitan cities in Turkey. Istanbul, the financial and the cultural capital of Turkey, has taken attention of the global investors with its dynamics and potentials. Consequently, state-owned industrial areas located in and around the city have become subject to privatization one by one. In this process, central government and local government have stimulated the transfer of the industrial areas by promoting them and by constituting action plans.

As a result of these attempts, many of the Turkish cities, mostly the metropolitan cities, have had to struggle with not only a physical segregations, but also economic, social and cultural changes occurred in a very rapid way without considering the possible outcomes/impacts of transformations.

3.1.4. Rapid urbanization

In Turkey, migration from rural to urban areas since 1950s has had a significant effect in the decentralization of industries and urban sprawl in metropolitan cities. At the early stages of this process, squatter houses (in Turkish literature *"gecekondus"*) appeared in Turkish metropolitan cities as the form of unplanned and uncontrolled urban development on the public lands around the industrial sites. These areas formed the settlements of workers which the industries require. By 1980, with the impact of technological improvements, liberal policies/ privatization, and globalization most of the industries moved to the organized industrial areas with the support of local and central governments. Many of the industries had preferred to abandon their properties in the inner cities due to the high land values and insufficient lands to expand. The decentralization of industries also gave way to the decentralization of *gecekondus* through the periphery of cities. As a result of this urbanization process, the former industrial areas in the city centers remained vacant with many contamination and other environmental problems.

In Turkish case, the urban regeneration-related issues have taken place by Habitat II meeting held in Istanbul in 1996 [15]. After this meeting, urban regeneration has become a new local policy in restructuring the cities through national plans and brownfields and *gecekondus* have seen as a tool for promoting cities in the global markets. However, these regeneration projects have put in force without comprehensive social, physical, and environmental sensitivities, generally far from a sustainable approach.

In recent years, most of the local authorities have renewed the upscale and detailed plans in order to galvanize the city's developments with new strategies and visions especially in metropolitan areas. Urban regeneration has been identified as the significant tool for achieving in the redevelopment of cities through global and local demands. In this process, brownfields have taken the attraction of developers and global investors with their locations and sizes. However, local authorities have tended to regard urban regeneration mostly as a project-based developments rather than a holistic restructuring process at the urban scale [39], and as a result, former spaces of industrial production sites can turned into luxurious residences, gentrified neighborhoods, office towers, shopping complexes, and the like [40]. Although these transformations negatively criticized by academicians and experts on the local agenda, local governors strongly support them in order to gain revenues and to integrate global cities network.

This situation is defined as "market-oriented transformation through governmental assistance" in many academic writings and reports [15, 39, 41–43].

3.1.5. Preservation policies

The concept of brownfield regeneration was taken place at the Turkish national agenda with the acceptance of the ICOMOS Montreal Action Plan by the Turkish National Committee of ICOMOS in 2001, which allowed industrial buildings to be conserved as part of Turkey's twentieth century cultural and architectural heritage [15]. Since the midst of 1980s, the potential of old industrial areas as a cultural heritage have recognized, and some projects were developed through the international preservation policies, especially in Istanbul. However, the sustainability issue neglected in most of those implementations. This is due in part to a general lack of awareness and an absence of nationwide debate about sustainable development. Furthermore, there was no participatory element in the decision-making, planning, and implementation of these brownfield regeneration projects [15].

Another international organization, DOCOMOMO, played a crucial role in fostering interest in conservation of built heritage which reflects the spirit of the Maschine Age in the world. The fact that many of the industrial modern buildings not being considered as the elements of heritage caused the demolishing or changing their original structure. In Turkey, DOCOMOMOs activities not only increase the attention to the industrial buildings as masterpieces, but also accelerate the institutional and academic interest on brownfield sites since 2002.

3.2. Barriers and assets in sustainable brownfields redevelopments in Turkey

In Turkey, the central and local governments' policies do not include the brownfield redevelopment related issues and on the other hand, the policies about sustainability devoid of the practice. Therefore, several obstacles occur in sustainable brownfield redevelopment, which are delaying the process.

The lack of liability and contamination issues in the environmental and land-use policies creates the major handicaps in the brownfield redevelopment process in Turkey. Despite some positive attempts for the integration of legislation to EU standards, there is still uncertainty related with liability and interventions to the contaminated lands.

The lack of nation-wide and city-wide database system is another vital obstacle in brownfield redevelopment in the country. It is nearly impossible for a researcher to determine the amount of brownfields and the level of their importance in the urban development. However, as a country whose metropolitan cities struggle with rapid urbanization and whose environmental resources are under the risk of deterioration, it is essential for Turkey to know the capacity of brownfields and to estimate the possible settlements for increasing population.

Due to the fact that most of the state-owned industrial sites are located at the centers of the cities on very strategic and valuable lands, their transformation and purchasing have been subject to many political and public debates because of the lack of transparency in the privatization process. These debates broadly interrupt the regeneration process and sometimes make the land remain idle for years.

Another essential issue in brownfields redevelopment is the effective involvement of different stakeholders in the process of redevelopment. Due to the lack of legal arrangements and policies for the participation of the citizens, environmental organizations, real estate developers, non-governmental organizations (NGOs), and local and central government agencies, the redevelopments of the brownfields, especially the ones located on valuable properties in the inner parts of cities, have been subject to speculations in the public agenda.

Despite the obstacles in brownfield redevelopment and preservation of historical industrial buildings, some good practices in metropolitan cities can provide some benefits in terms of sustainable development. Especially, in terms of preservation of the sites as the symbols of the cities' pervious identity, there is an increasing awareness at local and central levels. Although there is still need a comprehensive approach in handling the problem together with its social, economic, and environmental dimensions, these attempts can be considered as the signs of more sustainable solutions.

The SWOT analysis in **Table 1** summarizes the positive and negative aspects of sustainable brownfield redevelopment issue in Turkey with a strategic point of view.

	Strengths	Weaknesses	Opportunities	Threats
Environmental/ physical	-Existence of infrastructure -Accessibility -Historical heritage	-Lack of public awareness -Lack of legal arrangements related with environmental contamination -Lack of standards about remediation issues -Uncertainty in responsibilities	-Increasing public awareness in environmental related issues all around the world -Existence inside the city centres	-Urban sprawl -Greenfields -Lack of a definite environmental policy -Privatization policies -Lack of specific legal arrangements related with brownfields
Economic	-Valuable properties in the inner cities	-Handicaps in the transformation of state-owned properties into private properties -Financial risks due to contamination	-High market demand -High land prices	-Speculative implementations -Lack of administrative investments
Social		-Insufficient awareness in historical heritage preservation -Weak environmental groups		-Lack/insufficiency of public participation -Public resistance -Lack of collaboration between the stakeholders

Table 1. SWOT analysis for sustainable brownfield redevelopment in Turkey.

4. Recommendations for Turkey

In Turkey, brownfield redevelopment issue needs to be handled in a comprehensive manner beginning from policy formulation to determining design criteria. Cities in Turkey must struggle with many economic, social, environmental, and politic problems. At this point, brownfields represent opportunities in order to obtain sustainability and increase the living standards especially in urban areas.

One study of several EPA Brownfield Pilot projects found that for the sustainable brownfield redevelopment, the most common scenario involves a local government entity taking ownership of the site, characterizing environmental contamination, and then marketing the property to private sector [44]. This strategy is also can be very useful for Turkey. The Pittsburgh, USA experiences represent good practices to the essential roles of local authorities in brownfield redevelopments. In Pittsburgh case, URA did not only take place as a developer but also as facilitator that eliminate financial, social, and environmental obstacles [45]. The local governments can undertake the URA's role in small and medium sized cities in Turkey. On the other hand, a new body as a coordinator and technical assistant can be established in metropolitan cities in order to prevent speculative property transformations and to provide collaboration between central and local entities in financial and legal issues. Such an institution can also stimulate the public participation by organizing training programs. The structure of the institution should be generated with the interdisciplinary approach in a way to include environmental experts, urban planners, urban designers, economists, sociologists, architects, and lawyers.

In Turkey, the responsibilities for the waste managements are distributed between several institutions and organizations. The new authority can also coordinate and undertake some of their responsibilities in technical, financial, and monitoring issues.

As mentioned above, lack of transparency has been one of the major barriers in the redevelopment of brownfields. In Turkey, corruption in municipalities especially in urban development-related issues has been on the public agenda for decades [46]. Plan modifications and implementations generally have been subject to court cases. The transformations in state-owned lands, therefore, must be realized in transparency with public participation in order to eliminate speculations.

A big amount of brownfields in Turkey are originated from formerly state-owned firms. In order to turn those properties into marketable products, some essential changes should be done in the scope of urban development policies. At the same time, local and central authorities could prepare concrete measures to benefit brownfield redevelopment such as tax reduction and promotion of brownfields as equally interesting entities as greenfields.

Environmental taxes are widely used policy tool in many industrialized nations. Turkey also employs tax policy for environmental protection purposes. However, environment-specific taxes have been introduced primarily for revenue raising purposes. Since additional revenues collected through environmental taxes have not been allocated for environmentally friendly investments or services, their impact on the environment remains limited to gains from reduced consumption [37]. The legal arrangements should be done to provide the use of environmental taxes for environmental issues.

As brownfields are seen in the context of privatization, the legal and institutional framework should also be rearranged towards dealing with the remaining state property and enterprises in which the government still has responsibility. The remaining state-owned lands in metropolitan areas and small sized cities must be seen as opportunities for eliminating the environmental, economic, and social problems with a sustainable approach.

Another essential issue in Turkish case is the necessity of available data for the potential brownfields at national and local level. The absence of such inventories makes it impossible to estimate the dimension of brownfield problems and their possible socioeconomic costs. In United States and many of the Western countries governors try to determine the amount of brownfields. The data provide the vital knowledge for federal government in distribution of the public funds between the local authorities. As a country which struggle with unemployment and lack of financial sources, it is necessary to take the advantages of international funds for the projects where market-demand is low. Especially, the brownfields in the small and medium sized cities EU funds address opportunities for obtaining sustainable urban development and creating jobs on the brownfield areas. EU generally supports the well-defined and well-organized projects which aims environmental protection and eliminated air, soil, and water pollutions [47].

Among all, ensuring the community involvement is one of the cornerstones of brownfield redevelopments [48–51]. The impacts of brownfields can extend beyond the boundaries of the property to affect the neighborhoods and environment. Safety problems, environmental, and public health threats are some of their negative impacts on the surrounding communities. On the other side, their redevelopment can provide some economic and spatial opportunities. The increased efficiency of community-based organizations and community development corporations has triggered the governments to search for the ways to stimulate their participation [52]. Most of the governments recognized the fact that without their participation, it is nearly impossible to achieve a successful project in a reasonable time. However, in Turkey, most of the urban regeneration projects were realized without community participation. Therefore, there are still some challenges in the society against the regeneration projects.

All in all, the required attempts in order to achieve sustainable regeneration/redevelopment of brownfields at the national and local level in Turkey can be summarized as in the **Table 2**.

National level	Local level
Setting essential legal regulations	Designating brownfield sites (constituting an extended programmatic definition)
Setting a conceptual definition	Setting critical decisions about characteristics of the brownfield sites
Developing national, regional and city data bases of sites	Determining the goals and objectives of sustainable brownfield redevelopment
Determining national and regional development strategies	Determining the vision of cities through sustainability
Allocating national funds	Allocating local funds
Establishing transparency	Establishing transparency
Preparing national and regional development plans	Preparing master plans, detailed plans through sustainable development strategies
	Determining the possible stakeholders and their roles

Table 2. Required attempts for sustainable brownfield redevelopment in Turkey.

5. Conclusion

There has been an increase in the number of brownfields especially at the metropolitan areas as a result of the rise in population in Turkey. Turkey has been faced with migration from rural to urban areas since 1950s. In this respect, brownfield sites provide an alternative land-use supply, and thus a more sustainable way to preserve land resources. However, the most important issue of brownfield redevelopment, the contamination, has not been fully understood nor well recognized in the Turkish brownfield policies and strategies. In Turkey, there has been a tendency for planners and architects to handle the brownfields as a symbol of industrial heritage and as industrial archaeology areas, and therefore, the policies and implementations have been developed through the concept of preservation and urban regeneration rather than environmental clean-up.

It may be interesting to indicate that brownfields are formed not only as a result of rapid/ uncontrolled urbanization but also as a result of planning decisions in Turkey. Recent researches about the industrial movements in Turkish metropolitan cities [23, 24] have showed that most of the industrial firms changed their locations through master plan decisions prepared by local and central governments. Another significant factor in the formation of brownfields is the privatization of state-owned industries as a result of liberalization policies after 1980s. Most of the industrial areas and plants which were founded at the beginning of the Republican period have been subject to put up for sale through market demands. Establishing at the inner parts of the cities at the most valuable lands have not only increased the market demand but also the speculative sales due to the lack of transparency and lack of definite public policies for the redevelopment of these lands. Under these circumstances, brownfield redevelopments are often realized in a project-oriented manner, and this approach fails to meet the requirements of sustainable brownfield redevelopments.

The obstacles in brownfield redevelopment in Turkey broadly relate with indefinite policies at the national and local level, lack of specific brownfield-related regulations, lack of database and definition, insufficiency of practical knowledge and public participation, and lack of transparency in property transformations. There is no doubt that the environmental legislation in Turkey should require additional regulations concerned with contamination, remediation, liability, and public participation issues. The Western countries' experiences demonstrate that without a specific legal arrangement at the federal and state level, it is not possible to develop appropriate solutions for the sustainable redevelopment of brownfields [30, 45]. In Turkey, the responsibilities of central and local governments should be identified with the emphasis of the involvement of all the stakeholders.

Turkey, as a late industrialized country, can contribute the sustainable development strategies in brownfield redevelopments. Turkey has been at the point of development and environment dilemma for decades which constitute a big barrier in sustainability and sustainable urban regeneration issues. In the economically depressed areas, it is very difficult for the planners and politicians to support the policies considering environmental, recreational, historic preservation, and design principles and related solutions.

In Turkey, the planners and environmentalists have competed with political interest (often unsuccessfully) to protect the historical and environmental sources in the cities. On the other hand, the capitalism also reshaped the landscape with the emphasis of productivity more strongly, rather than sustainability and urban efficiency concerns.

Consequently, beginning with recognized gaps in brownfield-related literature and practice, this study has taken the picture of the sustainable brownfield redevelopment issues in Turkey and tried to develop some recommendations for future attempts through developed countries' perspectives.

Author details

Gökçen Kılınç Ürkmez

Address all correspondence to: gokcenkilinc@yahoo.com

Bursa Technical University, City and Regional Planning Department, Bursa, Turkey

References

[1] Dixon, T, (2006) "Integrating Sustainability into Brownfield Regeneration: Rhetoric or Reality? – An Analysis of the UK Development Industry" Journal of Property Research, 23(3) 237–267

[2] WCED (World Commission on Environment and Development), 1987. *Our Common Future*. Oxford: Oxford University Press.

[3] Mehdipour,A.; Nia, Hoda R., (2013) "The Role of Brownfield Development in Sustainable Urban Regeneration" *Journal of Sustainable Development Studies*, 4(2):78–87

[4] Oliver, L., Ferber, U., Grimski, D. Millar, K. and Nathanail, P. "The Scale and Nature of European Brownfields" available at http://www.cabernet.org.uk/resourcefs/417.pdf

[5] EUGRIS(European Information System Soil and Groundwater),(2011) Available at http://www.eugris.info/FurtherDescription.asp?Ca=1&Cy=9&T=Brownfields&e=93, (February 15, 2016)

[6] Guglielmi, Andrew O, (2006) "Recreating the Western City in a Post-Industrialized World: European Brownfield Policy and an American Comparison" *Buffalo Law Review*, 53 (4):1273–1312

[7] Nathanail, C. P. (2011) "Sustainable Brownfield Regeneration" in *Dealing With Contaminated Lands: From Theory towards Practical Application*, Frank A. Swartjes (ed.), Springer, London, New York

[8] Cengizkan, A. (2004) "Silahtarağa:Bir Endüstri Arkeolojisinin Anıtı" *Arrademento Mimarlık*, 171: 96–102

[9] Asiliskender, B. (2008) "Modernity and Housing; Spatial and Social Change in Kayseri By Industrial Enterprises of Turkish Republic", *Unpublished PhD Dissertation*, ITU Institute of Science, Istanbul.

[10] Asiliskender, B. (2010) "Cumhuriyet Sonrası Kayseri'de Modernleşme: Mekansal ve Toplumsal Değişim", *ITU Dergisi/a, Mimarlık, Planlama Tasarım*, 9(1):31–42

[11] Oral, E.O. and Ahunbay, Z. (2005) "Bursa'nın Ipekçilikle İlgili Endüstri Mirasının Korunmasi", *ITU Dergisi Mimarlık, Planlama, Tasarım*, 4 (2) :37–46

[12] Oğuz, D.; Saygı, H. and Akpınar, N. (2010) "Kentiçi Endüstri Alanlarının Dönüşümüne Bir Model: İzmit/Sekapark", *Coğrafi Bilimler Dergisi*, 8(2):157–167

[13] Cırık, U. (2005) "Ankara'nın İlk Endüstri Bölgesi-Kaybolan Tarih", Planlama, TMMOB Sehir Plancıları Odası Yayını, 2005/4, Vol. 34, 84–93

[14] Bostan, M., Erdoğanaras, F. and Gorer Tamer, N. (2010) "Ankara Metropolitan Alanı'nda İmalat Sanayinin Yer Değiştirme Süreci ve Özellikleri", *METU JFA*, 27(1): 81–102

[15] Cahantimur, A.I.; Öztürk, R.B.; Öztürk, A.C. (2010) "Securing Land for Urban Transformation Through Sustainable Brownfield Regeneration - The Case of Eskişehir, Turkey", *Environment and Urbanization*, 22(1):241–258

[16] Koçan, N. (2011) "Sanayi Alanlarının Dönüşümü: Uşak Eski Tabakhane Deri Sanayi Bölgesi" *The Black Sea Journal of Sciences*, 1(3):124–138

[17] Tolga, H.B. (2006) " Endüstriyel Alanların Dönüşümü, Kentsel Mekana Etkileri: Beykoz Kundura ve Deri Fabrikası İçin Bir Dönüşüm Senaryosu", *Unpublished Master Thesis*, Yildiz Teknik Univesity, Institute of Science, Istanbul.

[18] Yerliyurt, B. and Aysu, E. (2008) "Kentsel Kıyı Alanlarında Yer Alan Sanayi Bölgelerinde Dönüşüm Potansiyelinin Değerlendirilmesi: Haliç-Tersaneler Bölgesi", *Megaron YTU Mim. Fak. E-Dergisi*, 3(2):194–205

[19] Köksal, T.G. (2005) " Istanbul'daki Endüstri Mirası İçin Koruma ve Yeniden Kullanım Önerileri", *Unpublished PhD. Dissertation*, ITU Institute of Science, Istanbul

[20] Köksal, T. G., (2001) "Istanbul'un Önemli Bir Endüstri Arkeolojisi, Haliç Tersaneleri", *Tarih Vakfı* , 27–32, İstanbul,.

[21] Akın, N., Salman, S. G. and Kahya, Y. (2002) "20. Yüzyıl Endüstri Mirasının Korunmasına Bir Örnek: Bakırköy İspirtohane Binası", *Arredamento Mimarlık*, sayı: 5, s.110–116.

[22] Tanyeli, G., (2003) "Istanbul'da Endüstri Arkeolojisi, İki Su Tesisi Örneği, *İstanbul Dergisi*, 46: 94–98.

[23] Kıraç, B. (2001) " Türkiye'deki Sanayi Yapılarının Günümüz Koşullarına Göre Yeniden Değerlendirilmesi Konusunda Bir Yöntem Araştırması" *Unpublished Phd Dissertation*, MSU Institute of Science, Istanbul

[24] Oral. A.H. (2006) "İşlevini Yitirmiş Endüstriyel Alanların Dönüşümü İçin Bütüncül Yaklaşım: Haliç Yerleşim Örneği", *Unpublished Master Thesis*, Gebze Istitute of Technology, Institute of Engineering and Science, Gebze, Kocaeli.

[25] Gedikli, A.P. (2002) "The transformation of Urban Waterfronts in Istanbul and Evaluation of the Golden Horn Arsenals" *Unpublished Msc Thesis*, METU, Ankara

[26] Kazas, J. (2008) "Endüstriyel Miras Kapsamındaki Alanların Kentsel Yenilemeyi Oluşturmadaki Rolünün Ördelenmesi: Ödemiş Örneği", *Unpublished PhD Dissertation*, YTU Institute of Science, Istanbul

[27] Polat, S.; Erbil, Y. (2011) "Trademark City Bursa: Merinos Park And Ataturk Culture and Congress Center", *Ozean Journal of Applied Sciences*, 4(2):171–181

[28] Türer Başkaya, A. (2010) "Ways to sustainable brownfield regeneration in Istanbul" *ITU A|Z*, 2010- 7 / 2 – A, pp.74–88

[29] Ergen, M. (2013) "A New Methodology Proposal For Urban Regeneration of Brownfield Areas; Case Study Of Zonguldak City, Turkey" *Unpublished Ph.D. Dissertation*, Faculty of Spatial Planning of Dortmund University of Technology.

[30] Kılınç G. (2011). "Brownfield Redevelopment in Turkey: Implementation Perspectives Learned from US Experience", *SECoPA 2011- Building Trust and Confidence in the Profession of Public Administration Conference*, September 21–24, 2011, New Orleans, LA

[31] UNDPTURKIYE (1999) "National Environment Action Plan of Turkey", Available at https://issuu.com/undpturkiye/docs/neap-turkey (March 1,2016)

[32] UNESCAP, (1999) National Environmental Action Plan of Turkey, available at http://www.unescap.org/stat/envstat/neap-turkey.pdf (August 31, 2011)

[33] Kılınç, G. (2012a) "The Impact of Environmental Movements on the Development of Brownfield Revitalization Policies in U.S. and Recommendations for Turkey" 2nd International Congress on Urban and Environmental Issues, (4–6 May 2012), *Papers and Proceedings of Congress*, S.394–409 Trabzon, Turkey

[34] İzci, R. (2005) "The Impact of European Union on Environmental Policy" in *Environmentalism in Turkey: Between Democracy and Development?*, F. Adaman, M. Arsel (eds.) Ashgate, Burlington, USA, 87–100

[35] Budak, S. (2000) Avrupa Birligi ve Turk Cevre Politikasi. Avrupa Toplulugunun Cevre Politikasi ve Turkiye'nin Uyum Sorunu, Buke, Istanbul.

[36] Ekmeztoglou, T., Baladimos A., and Budak S. (2001) Türkiye'nin Avrupa Birliğine Katılım Süreci: Avrupa Birliği'nin Çevre Politikasi ve Türkiye'nin Uyumu, Kitap 5, H. Cansevdi (ed.) IKV, İstanbul.

[37] Kalaycıoğlu, S. ; Gönel, F. (2005) "The Role of Business in Environmental Policy Making" in *Environmentalism in Turkey: Between Democracy and Development?*, F. Adaman, M. Arsel (eds.) Ashgate, Burlington, USA, 117–130

[38] Tecer, M. (1992) "Privatization in Turkey" Availabe at http://www.ceri-sciences-po.com/publica/cemoti/textes14/tecer.pdf (September 14, 2011)

[39] Güzey, O (2009), "Urban Regeneration and Increased Competitive Power: Ankara in an Era of Globalization", Cities, 26(1): 27–37.

[40] Bezmez, D. (2009) "The Politics of Urban Waterfront Regeneration: The Case of Haliç (the Golden Horn), Istanbul", International Journal of Urban and Regional Research, 32(4):815–840

[41] Uzun, C. N. (2006), "Yeni Yasal Düzenlemeler ve Kentsel Dönüşüme Etkileri", *Planlama*, Chamber of Urban Designers Publishing, Ankara, Vol.2 pp. 49–53

[42] Kurtuluş, H. (2006), "Kentsel Dönüşüme Modern Kent Mitinin Çöküsü Çerçevesinden Bakmak", Planlama, Chamber of Urban Designers Publishing, Ankara, Vol.2, pp. 7–12

[43] TMMOB Chamber of City Planners Board (2006), "Kentsel Dönüşüm Değil Rant Amaçlı Tasfiye Yasasi", *Planlama*, Chamber of Urban Designers Publishing, Ankara, Vol. 2, pp.5–6

[44] USEPA (1998) Final Guidance for Incorporating Environmental Justice Concerns in EPA's NEPA Compliance Analyses. Washington, DC

[45] Kılınç, G. (2012b) "Assessment of Brownfields Redevelopment Projects in Pittsburgh, PA with their Economic, Social and Institutional Dimensions", *AESOP'2012*, (10–14 September 2012), Ankara, Turkey.

[46] Kılınç, G., Özgür, H., Genç, F. N., (2012). "The Possible Sources of Ethical Issues in Urban/Physical Planning in Turkey", Turkish Studies, Vol. 13, No.1, 41-, http://dx.doi.org/10.1080/14683849.2012.653126

[47] TEPAV (2015) Turkey's Compliance with Industrial Emissions Directive, Available at http://www.tepav.org.tr/upload/files/haber/1427475571-5.Turkey___s_Compliance_with_the_Industrial_Emissions_Directive.pdf (March 1 ,2016)

[48] Brachman, L. (2004) "Turning Brownfields into Community Assets: Barriers to Redevelopment" in *Recycling the City : The Use and Reuse of Urban Land*, Rosalind Greenstein and Yesim Sungu Eryilmaz (eds.) Lincoln Institute of Land Policy, Cambridge, Massachusetts, 67–88

[49] Dewar, M & Deitrick S. (2004) "The Role of Community Development Corporations in Brownfield Redevelopment" in *Recycling the City : The Use and Reuse of Urban Land*, Rosalind Greenstein & Yesim Sungu Eryilmaz (eds.) Lincoln Institute of Land Policy, Cambridge, Massachusetts, 159–176

[50] McCarthy, L. (2009) "Off the Mark? : Efficiency in Targeting the Most Marketable Sites Rather Than Equity in Public Assistance for Brownfield Redevelopment", Economic Development Quarterly, 23(3), 211–228

[51] Wernsted, K., Hersch, R. (1998) 'Through a lens darkly- Superfund spectacles on Public participation at Brownfield Sites', 9 Risk: Health, Safety & Environment,9:153–173

[52] APA (American Planning Association) (2011), Reuse: Creating Community-Based Brownfields Redevelopment Strategies, Available at http://www.planning.org/research/brownfields/pdf/brownfieldsguide.pdf (18 August 2011)

2

Smart Specialisation Strategies as Drivers for (Smart) Sustainable Urban Development

Claudia Trillo

Abstract

The *place-based* approach is at the forefront of the current European cohesion policy reform. The *Smart Specialisation Strategy (S3)* boosts economic development through research and innovation. It is built on *place-based* areas showing higher potential for growth. Although S3 is deeply rooted in the uniqueness of the place, its implementation is still far from being clearly spatial-led. The research programme MAPS-LED (Multidisciplinary Approach to Plan Smart specialisation strategies for Local Economic Development), financed by RISE-Horizon 2020, aims at filling this gap. This chapter discusses a novel concept developed with the aim to incorporate a spatial component in the innovative urban clusters, e.g. *the urban pattern as cognitive infrastructure for S3*. The preliminary findings from the case study of Kendall Square, a former brownfield located in Cambridge (MA), in close proximity with the Massachusetts Institute of Technology, led to conclude that it is essential to complement S3 implementation with spatial interventions in the built environment to create a *physical ecosystem* supportive of innovation, including: shared facilities and private small businesses facilitating interaction; public services and facilities that allow preserving uniqueness of the place and inclusiveness. Furthermore, the spatial pattern has to be supportive of a walkable and dense environment.

Keywords: smart specialisation strategies (S3), sustainable urban development, innovation districts, social innovation, Kendall Square, MAPS-LED, urban pattern as cognitive infrastructure

1. Introduction

Policy makers, planners, stakeholders at all latitudes constantly face the issue of developing and implementing better policies to support local economic development and offer people better

living environments and well-being. How it is made, varies a lot. Nowadays, the European Commission is undertaking the huge effort of launching something that explicitly implies risk-taking and therefore particularly needs and evidence-base for actions. The research role is to reflect on this challenge from a theoretical perspective that might possibly suggest paths and solutions.

This chapter stems from a broader research project financed by the European Commission and aimed at approaching the issue of strategy building, developing and implementation from the perspective of architects and planners, temporarily contaminating themselves with economic matters to try to bridge the gap between shape of the built environment and economic growth. The Smart Specialisation Strategy (S3) will be discussed through a spatial-oriented perspective, arguing that if *place* matters, then also *space* matters, with all the implications within a concept that incorporates environmental behavioural science, ecosystem and social values, cultural assets and identity. All these elements are absolutely essential in a *sustainability* perspective. Therefore, the overall discussion has in the background the *fil rouge* of demonstrating how sustainable development (environmental, social and economic) can be systematically embedded in S3, in particular, in the urban built environment.

This chapter briefly explains what Smart Specialisation Strategy is and then discusses it in a critical perspective, by clarifying explicit and less evident theoretical legacy of this rationale, with the aim to support the construction of a robust logical framework suitable to produce further novel approaches. Because of the dynamic nature of the topic, even the theoretical section, rather than relying on a review of the literature, is nurtured by up to date interviews. The chapter includes a discussion of the research hypothesis through empirical data gathered in a US case study, Kendall Square. The major expected impact of this research is the opportunity to support current implementation of S3 policies in Europe, both in competitive and in lagging behind regions. To reinforce the transferability of the provisional findings, the field work in the States has been preceded by some exploratory investigation in Europe, aimed at observing the current gaps to be filled on the basis of the gap analysis of extant S3. The current state of S3 in the Greater Manchester Area (Northern England) and in the Calabria Region (Southern Italy) has been analysed, both on desk and through a set of informal unstructured discovery interviews with key stakeholders, in order to find out weaknesses and potentials. This preliminary analysis showed that both in lagging behind regions and in competitive regions gaps in the current S3 still exist, and that a spatial-led approach could be supportive in filling them. Therefore, although at a preliminary stage, the conclusions in this chapter may be of interest for European planners, policy makers and stakeholders looking for effective implementation of S3 in a spatially- oriented perspective.

2. Setting the overall policy framework for S3

The Smart Specialisation Strategy (S3) is an essential component of the current Europe 2020 Strategy, seeking to bring Europe towards a smarter, more inclusive and sustainable growth [1]. In particular, S3 is a strategy for economic development that targets research and innovation and involves an incremental approach based on the development of a shared vision. It is built on place-based areas showing higher potential for growth and needs to be developed by involving multiple stakeholders with a strong commitment to prioritise in a knowledge-led perspective, not necessarily focusing on the high-tech sector, nor on sectors that are already strong [2].

S3 has been introduced in the late 2000s as the main result of the work conducted by the Research Commissioner Janez Potočnik's expert group, also known as the Knowledge for Growth (K4G) expert group, founded by the European Commissioner in March 2005 with the task to address the issue of embedding innovation for promoting growth within the European Member States, legacy from the Lisbon strategy. Not only the European Commission, but also other organisations such as the OECD are highly interested in this innovative approach [3] that has been recently systematised in the literature [4, 5]. A key concept underpinned in the Smart Specialisation is the importance of knowledge, not meant as a mere technicality, rather than as *embedded knowledge* among the actors of the economic ecosystem. Stemming from this position, the methodology for developing appropriate strategies rooted in embedded knowledge could not be anything different from an ascending, bottom-up approach, characterised by *discovery* and *risk-taking*, and finally, leading to something *unique*. As clarified by Foray et al.: "It should be understood at the outset that the idea of smart specialisation does not call for imposing specialisation through some form of top-down industrial policy that is directed in accord with a pre-conceived grand plan" [6]. It is suggested that, rather than relying on external consultants to develop an abstract strategy, policy makers should elicit a process of entrepreneurial discovery, in order to find out what are the assets, even still hidden, that could be worth supporting. These latter could be niches of excellence, and the process of discovery in itself should act as an activator to unleash their potentials by making entrepreneurial actors "play leading roles in discovering promising areas of future specialisation, not least because the needed adaptations to local skills, materials, environmental conditions, and market access conditions are unlikely to be able to draw on codified, publicly shared knowledge, and instead will entail gathering localised information and the formation of social capital assets" [6].

Moving forward, the S3, also named Research and Innovation Strategies for Smart Specialisation (RIS3), is now "a key part of the proposed EU Cohesion Policy reform supporting thematic concentration and reinforcing strategic programming and performance orientation" [7].

By overcoming a one-size-fits-all approach, "the RIS3 requires an integrated and place-based approach to policy design and delivery. Policies must be tailored to the local context, acknowledging that there are different pathways for regional innovation and development" [7].

A key aspect of smart specialisation is the emphasis on the principle of prioritisation in a *vertical* logic—to favour some technologies, fields, population of firms—*non-neutral*. Foray and Goenaga suggest to summarise the principles of S3 as follows: (1) Granularity, that is, the level should not be too high; (2) Entrepreneurial discovery, with entrepreneurs -in the broadest sense—who discover, produce information and transform the activities; (3) Priorities will not be supported forever; (4) S3 is an inclusive strategy; (5) S3 has experimental nature and risk taking is needed [8].

The legal basis for incorporating the RIS3 within the current programmes is provided by the Regulation (EU) 1301/2013 of the European Parliament and of the Council of 17 December 2013. The implementation of RIS3 across the EU has to be ensured by managing authorities through dedicated and mandatory policy frameworks. EU Member States and regions are required to have S3 in place according to the RIS3 ex-ante *conditionality*, that is, a compulsory requirement that if not met in the agreed timeframe, prevents managing authorities from keeping on spending the given EU funds. Support in putting the RIS3 forward is offered by the European Commission particularly through a specific tool, the S3 Platform [9].

Influences on the construction of the conceptual framework of the S3 can be found in several theoretical positions and theories. On the basis of the industrial Italian experience, the concept of *industrial district* was developed in the 1980s by Becattini [10, 11], drawing from the Marshall agglomeration theory [12]. In particular, he considered the local community as a sort of social glue suitable to produce economic added value. In this approach, the seeds of the communitarian root of the concept of *embeddedness* were planted [13]. It will take about two decades for them to fully blossom.

With less emphasis on the social component of the proximity, and more attention for the scale advantages, in the 1990s, Porter developed the concept of *cluster*, defined as: "A geographically proximate group of interconnected companies, suppliers, service providers and associated institutions in a particular field linked by externalities of various types" [14]. As possible examples, Porter mentioned the financial services cluster in New York City, the cluster producing medical device developed in the Boston area, and the IT clusters existing in Texas (Austin) and in the Silicon Valley. The cluster rationale was deeply intertwined with the concept of competitiveness [15].

Building on this concept, recognising the importance of the cluster structure in the US economy, huge and systematic efforts have been done even at institutional level to pursue a reliable and shared knowledge on cluster dynamics, leading to the construction of a dedicated platform, such as the Cluster mapping platform: "The U.S. Cluster Mapping website is a national initiative that provides open data on regional clusters and economies to support U.S. business, innovation and policy, (where) users will find interactive, robust data and tools to understand clusters and regional business environments, improve institutions, and locate appropriate partners across the country" [16].

The relevance of clusters to the US economic success and the political awareness on the significance of this topic clearly emerge, while analysing the data contained in the platform. Clusters, far from being a theoretical concept, have become a conceptual framework to coordinate and even further activate all scales of *clusterizable* initiatives, encompassing national, regional and local stakeholders, entrepreneurs, companies, associations. In theory, the potential underpinned in the US platform is that the richness of details creates an outstanding opportunity not only for advancing in terms of knowledge, but also for supporting further networks and, finally, the *entrepreneurial discovery* that S3 is seeking to promote. In practice, the impact of the US platform on reinforcing clusters can be further exploited [17].

Following the work conducted by the Department for Competitiveness in Harvard, while on one side of the Atlantic the US Department of Commerce, Economic Development Administration was turning an academic platform into an official public initiative by funding the Clustermapping platform as a federal programme, on the other side of the Ocean also the European Commission decided to introduce an analogous platform, namely the Cluster European Observatory, whose architecture is similar to the US prototype [18]. In Europe, the dataset made available through the Cluster Observatory platform is coupled by another platform that targets companies and is specifically aimed at eliciting clusters reinforcement and further development, namely the ECCP (European Cluster Collaboration Platform) [19].

The common goal of the US and EU platforms is not only to build a structured knowledge on the cluster policy both in the States and in Europe, but also to create opportunities for making cluster work in a rationale embedding shared knowledge and entrepreneurial discovery as major triggers. This is a common trait that makes cluster policy useful to support successful S3 implementation, behind the simple network rationale.

The notion of *entrepreneurial discovery* was introduced by Haussmann and Rodrik [20] as a *self-discovery process* and is constantly recalled by Foray and Goenaga [21], which clearly mention the legacy of the New Industrial Economy approach in discussing the above mentioned five principles of S3. This core feature of S3 leads to another key concept at the forefront of current European strategies, that is, *social innovation*. A strong link exists between the S3 strategy, the cluster policy and the concept of social innovation as developed by the European Commission, a cross-cutting approach suitable to be implemented as trans-sectoral innovation. In the Guide to Social Innovation—commissioned by DG Regional and Urban Policy and completed with DG Employment, Social Affairs and Inclusion with inputs by various other Directorates General (such as, among others, DG Enterprise and Industry and DG Research)—social innovation is defined as: "the development and implementation of new ideas (products, services and models) to meet social needs and create new social relationships or collaborations. It represents new responses to pressing social demands, which affect the process of social interactions. It is aimed at improving human well-being. Social innovations are innovations that are social in both their ends and their means. They are innovations that are not only good for society but also enhance individuals' capacity to act" [22].

Social Innovation is deeply intertwined with socially- oriented and citizens- led urban regeneration and entails the granularity of the scale where it is more likely to happen through catalysts such as shared knowledge and innovation building. Cross-fertilisation between S3 and Social Innovation can be pursued at the strategy implementation stage. The Guide suggests that Social Innovation can be included in the Smart Specialisation Strategy and Plan and /or incorporated in Social Innovation Clusters/Park (Step 4 and Step 8) [22].

The concept of social innovation may be supportive when seeking to understand some features in S3, that still lack a clear spatial definition. In particular, if the concept of *granularity*, and in particular of *spatial proximity*, is precondition to achieve cross-fertilisation across ideas and expertise, as advocated by almost all the examples suggested as good practice in the guide, what it the *metric* of proximity? Is the proximity needed for enacting social innovation processes the same scale of proximity necessary to activate effective clusters or some specific kind of clusters, perhaps those that are more relying on innovation? Can we measure this proximity?

If cooperation is based on mutual trust, personal knowledge and social reputation, in some cases the scale of proximity requested for activating successful clusters overlaps with the scale of the proximity necessary to enable successful social activation processes. Cluster theory mainly rests on the opportunity for up-scaling economic mechanisms, thus creating advantages for the participating companies, and, in addition to it, also on shared knowledge and exchange of competences within a given network, while this latter is central in social innovation mechanisms and in S3. In facts, the geography of clusters overlaps with the labour markets, and—typically—cluster analysis and clusters data gathering are conducted at a regional scale. The innovation component, essential in the social innovation process and in S3, can be optional in clusters, ideally—but not *necessarily* innovative. These and other similarities and differences are systematically discussed in a recent report commissioned by the DG Research of the European Commission and produced by a group on independent experts chaired by Ketels. In particular, the most important differences between clusters and S3 follow: "S3 focuses on specific innovation-intensive sectors while clusters apply to a broader set of sectors in the economy. S3 aims to exploit emerging linkages between economic activities that can cut across traditional cluster boundaries.... The explicit goal of cluster policies is often to enhance the performance of existing clusters(...). Clusters are potential elements of a regional innovation eco-system, while S3 are wider policies aimed at transforming this eco-system. Clusters can come close to "smart specialisation domains" if they stimulate new types of knowledge spillovers with a high leverage effect on the growth path of the economy" [8].

Several attempts have been made in the cluster literature to find out a possible taxonomy; however, in knowledge-intensive clusters, the triple helix concept (referred to the relationship between universities, enterprises and government) is essential. As Porter has been highlighting since 1990, four intertwined factors concur to the creation of a competitive environment for companies, depicted in the form of a diamond. This combination works in two ways, since investing in public good, always seen as a typically public activity, becomes important for the private sector itself [23]. In an ecosystem approach, private *vs* public interests' boundaries finally blur. The same concept of producing social services as a matter of business is gaining growing interest in the private sector [24]. The multiple actors involved with different roles in supporting the economic growth depict the complexity of the entrepreneurial ecosystem, from which S3 should stem. As stated by Foray and Goenaga, those who are asked to promote S3 by discovering "the domains of R&D and innovation in which a region is likely to excel given its existing capabilities and productive assets", are "entrepreneurs in the broader sense (innovative firms, research leaders in higher education institutions, independent inventors and innovators)" [19]. As in S3, also in cluster policy the whole context matters.

3. S3: how far do place and space matter?

The importance of a site-specific and context-related approach is at the forefront of the current cohesion policy reform, since in 2009 the "Barca report" was released. Following an intense discussion, nurtured by three thematic hearings, one workshop and five policy seminars involving 80 both EU and non EU experts, this report clarifies that a possible failure in the European policies is due to a lack of *place-based* approach, thus advocating for the opposite, i.e.: "A place-based policy is a long-term strategy aimed at tackling persistent underutilisation of potential and reducing persistent social exclusion in specific places through external interventions and multilevel governance. It promotes the supply of integrated goods and services tailored to contexts, and it triggers institutional changes. In a place-based policy, public interventions rely on local knowledge and are verifiable and submitted to scrutiny" [25].

By supporting a territorial based approach, the "Barca report" suggests to ground the reform on pillars, consistent with the S3 approach, such as including the promotion of a learning process, of experimentalism, of mobilisation of local actors. The similarity between the locally-grounded approach of S3 and the place-based approach stemming from the Barca position, converging towards a *non-neutral* approach, has been highlighted by Foray [26]. The position expressed in the "Barca report" has been framed within the current debate between *spatially-blind vs place-based* approach in policy implementation. According to Barca et al. [27], a spatially-blind approach is that supported by the World Bank's report [28], that recommends to design policies without taking in consideration space, in order to ensure efficiency, equal opportunities and improvement of the life conditions, as well as it is spatially-blind the Sapir et al. report [29], that recommends to pay little attention to the sub-national scale. Under the second approach, the place-based one, in addition to the Barca Report and among others, it is possible to include in particular the OECD [30] position, that recommends a region- specific perspective capable to unleash assets and to exploit synergies.

More in depth [27], in the place-based approach it is essential to consider the interaction between institutions and geography in order to understand the best policy options for a given territory. Moreover, also the impacts of a policy should be assessed by considering those two factors, thus requiring tackling both the regional and the local context.

In terms of governance, the importance of a closer level of proximity to the local assets and knowledge leads to the inadequacy of the national scale in capturing appropriate policies, more specifically: "(...) by acknowledging the limits of the central state to design good local development policies, place-based strategies recognise the need for intervention based on partnerships between different levels of governance" [27].

Since the early documents on the spatial perspective of European policies, culminating in the 1999 European Spatial Development Perspective [31], the importance of a spatially-led perspective in European policies has been advocated from different authors since long time and the debate is still relevant [32, 33]. The spatial perspective is the physical setting for enabling place-based policies grounded in the specific territories. A lack of territoriality even interferes with a transparent exercise of democracy [34], thus, far from being a merely geographic concept, space and territory are real and proper enablers of context specific policies and related implementation. Moreover, because in the current EU programming period the concept of *territoriality* is embedded within important and innovative policy instruments, such as the Integrated Territorial Investments and the Community-Led Local Development, gaps in a place-based approach would undermine the effective implementation of new instruments holding a high potential of unleashing context specific assets. A better awareness of the governance within place-based S3 could support the creation of effective network of stakeholders for the Community Local Led Development strategies implementation, an innovative approach in the ESRF and ESF programmes implementation drawn from the LEADER approach and not yet fully developed outside the rural contexts. Despite of their strong roots in a place-based approach, S3 are still far from being clearly spatial-led strategies. This may depend on the original conceptualisation of S3, developed from a spatial idea [35]. It can be therefore problematic to translate them into genuine place-based policies, reflecting a consistent social innovation based institutional framework, particularly in those regions, still lacking in clear and updated spatial frameworks.

In order to fill this gap, a research programme has been proposed and accepted for grant under the Horizon 2020 programme, namely MAPS-LED (Multidisciplinary Approach to Plan Smart specialisation strategies for Local Economic Development) [36]. This program, run by a consortium of 6 universities in EU and in the US aims in particular at connecting three important key-factors including: (1) Governance—both in cluster policies and in terms of embeddedness; (2) Localization—as spatial and place-based approach; (3) Territorial network—as innovative milieu supporting social innovation, also based on urban-rural links. The project is building a novel methodology to assess and exploit the potential of different clusters, networks and chains in shaping spatially-led S3 policies for local economic development through a spatial-led approach. After having explored the potential of S3 both through spatial planning (city-region and S3) and regional economy (cluster policy, territorial milieu and S3), the project will develop and test a tailored evaluative tool suitable to capture the socio-economic spillovers of S3. By understanding how S3 can be translated and implemented into spatially-oriented local development policies, in line with the territorial agenda of Europe 2020 incorporating a place-based dimension, the expected results are: (1) to identify and examine S3 in terms of spatial, social and environmental factors; (2) to take into account local needs and opportunities driving regional policy interventions not only to emphasise "Key Enable Technologies", but also to empower local innovation process—tacit knowledge, embedded social networks, innovative milieu; (3) to build and test an evidence-based methodology for recognizing and assessing emerging and potential S3, corroborated by successful factors of existing clusters [36].

4. Urban patterns as cognitive infrastructure for successful S3

Shifting from the regional towards the urban scale, the concepts of social innovation, entrepreneurial discovery and local embeddedness can be found in the recent theorisation of *innovative district*. Starting from a metropolitan centred perspective, the importance of the scale and related *metric* is a recurrent concept for understanding the assets of the place: "The next economy must have four characteristics: higher exports, to take advantage of rising global demand; low-carbon technology, to lead the clean-energy revolution; innovation, to spur growth through ideas and their deployment; and greater opportunity, to reverse the troubling, decades-long rise in inequality. Metros will take the lead on all four fronts. Our open, innovative economy increasingly craves proximity and extols integration, which allows knowledge to be transferred easily between, within, and across clusters, firms, workers, and supporting institutions. The vanguard of these megatrends is largely found not at the city of metropolitan scale (...) but in smaller enclaves, which are increasingly being called innovation districts" [37].

The scale of the innovation districts is clear. They are walkable urban environments, typically featured following the current urban design tendency of creating vibrant spaces offering a variety of uses, shared places, and accessibility. If we look at an innovation district as the brain of an innovative cluster [38], it follows that urban patterns have to be treated as cognitive infrastructure of the collective knowledge production.

What is the rationale that creates value out of the proximity? Recent studies focus on the creation of successful groups of players capable of activating cooperation on the basis of the mutual trust [39]. Building on this concept, an extensive literature is blossoming in support of mutual trust and cooperation as triggers for successful social dynamics (for example [40]). What cluster potentially does is that it increases the roles of reputations by increasing the frequency of interaction and also how observable actions are. Therefore, innovative urban spaces, in order to be supportive for a specific kind of entrepreneurs, those who are willing to cooperate in producing shared knowledge, has to support density, accessibility, and also shared spaces that make good and cooperative actions *frequent and observable* [41].

Figure 1. MIT innovation cluster, March 2016 (Courtesy of the MIT).

The first attempt to corroborate this research hypothesis has been done through the investigation of the hidden mechanisms supporting the outstanding competitiveness of a US based innovative district, the Kendall Square in Cambridge, MA (**Figure 1**). The history of Kendall Square is intrinsically related with the presence of the Massachusetts Institute of Technology, which makes this case study, obviously, almost unique. However, because of this uniqueness, it can well explain how, even in an outstanding context in terms of innovation, still spatial factors play a significant role and are considered relevant both by public and private actors. Kendall Square is a former brownfield located in Cambridge (MA), opposite side of Charles River. It started in 1868 as an industrial district and consolidated this function with the opening of the first underground line nearby. The presence of the Massachusetts Institute of Technology dates back to 1916. Following the Second World War, the area entered an era of decline, which the Cambridge Redevelopment Authority (CRA), established in 1955, sought to reverse also through the clearance of 29 acres of land for the accommodation of NASA. Because of a change in the federal government strategies, the plan was no longer implemented, and the vacant land was partly redirected to the Department of Transportation. A shift in the approach to the redevelopment of the area, managed as a detached industrial estate, happened first with the implementation of the East Cambridge Riverfront Plan, then with the 2001 Citywide Rezoning. Walkability, quality of open spaces and mixed-use real estate became the norm in the area [42, 43]. Recent massive capital investments confirm the tendency to invest in the area with high quality interventions [44]. The importance of the urban structure as catalyst for local development is acknowledged both by public and private stakeholders [45, 46]. As emerges from the current planning documents (**Figure 2**), the connection between urban fabric and attractiveness of the area for private companies is evident: "A dynamic public realm connecting diverse choices for living, working, learning, and playing to inspire continued success of Cambridge's sustainable, globally-significant innovation community" [47].

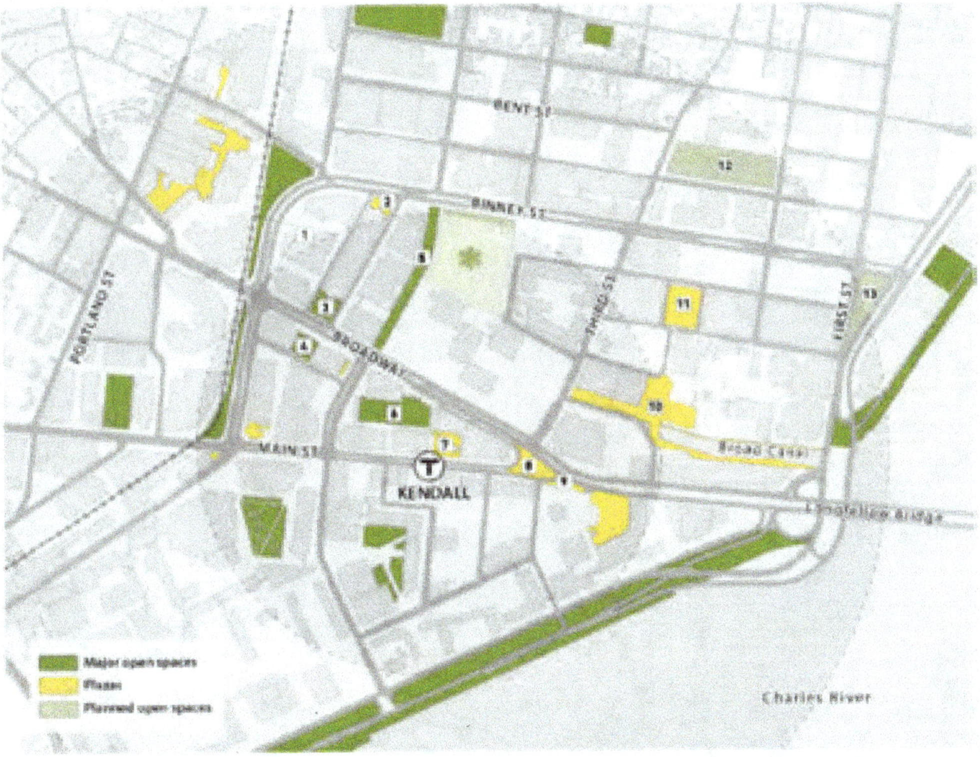

Figure 2. Kendall Square open spaces (Source: [42], p. 28).

Furthermore, recent studies on the companies' behaviour in this area proved how the cluster traditional policies based on subsidising "anchors" that—once settled down—would attract smaller companies, is now coupled by a bottom-up oriented perspective, in which a significant number of small, dynamic, highly innovative companies create the favourable ecosystem for attracting big companies, interested in having an interaction with young talents and possibly in incorporating smaller (and cheap) companies with higher potential for growth [48]. Therefore, public policy makers should also support the creation of a cluster of innovative and cutting-edge start-up companies, rather than seeking to attract a big one to make them follow. This creates the need for urban environments that are attractive, as Florida suggests, for young talents [49]. How much companies value the competitive advantage of being in this kind of environment is testified by what recently happened to a leading pharmaceutical company in the Kendall area. Biotech is one of the historical companies located in Kendall Square, founded by a MIT professor, Sharp, who at the time he launched the company, wanted to work close to his laboratory. Despite of its roots in the area, in recent years a controversial decision was made, to move the Biotech headquarters to the cheaper suburb of Weston. In a few years, this decision was questioned and the willingness to return to the area prevailed [50]. The reasons are clear: "Other biotech companies have come to the neighbourhood to take advantage of the healthy infrastructure in Cambridge and its vibrant bioscience community. While there were many individuals and organisations involved, MIT faculty members and administrators indeed played a major role in reviving Kendall Square, because they understood that in order to build a thriving bioscience programme, they would have to build a thriving community of talented people—at MIT and beyond" [51]. This view is shared by experts on cluster and industrial policies in the Massachusetts, who suggested that what brought Biotech back to the Kendall area was the "atmosphere" [52].

Figure 3. Open spaces and public services around the Kendall Square area (author's picture).

Figure 4. Companies located in the Kendall's immediate surroundings: Akamai (author's picture).

Private companies perceive the economic benefit of being localised in an innovative district, and are willing to pay the extra costs associated with a more expensive location in order to get extra benefits in return, including the well-being (and related increase of productivity) of their employees and the opportunity to benefit from the powerful network of informal and multi-disciplinary connections, made possible by the specific features of the urban fabric (**Figures 3–12**).

Figure 5. Companies located in the Kendall's immediate surroundings: Biogen (author's picture).

Figure 6. Restaurants, coffee shops, in the Kendall's surrounding (author's picture).

Figure 7. Amenities in the Kendall's immediate surroundings (author's picture).

Figure 8. Companies located in the Kendall's immediate surroundings: Genzine (author's picture).

Figure 9. Companies located in heart of Kendall: Microsoft (author's picture).

Figure 10. Bike sharing facilities in Kendall (author's picture).

Figure 11. Transit station, bicycles, pedestrians: a walkable environment (author's picture).

Figure 12. Large sidewalks, benches, bus stops: a pedestrian friendly place (author's picture).

Figure 13. Concierge in the Kendall Square CIC building (author's picture).

Besides the urban pattern encouraging knowledge and innovation building, Kendall Square also includes key-hotspots for informal decision making and cross-clustering, such as the Cambridge Innovation Centre, CIC [53]. The CIC is not an incubator neither an accelerator, it is a private entrepreneurial activity based on renting shared and flexible office spaces designed with an innovative rationale. It currently hosts over 700 companies across two buildings, located in Kendall Square and in downtown Boston, about 500 of which are start-ups. The

Kendall Square building includes the Venture Café, a sister non-profit with the mission of bringing together entrepreneurs, venture capitals and the greater Boston start-up community. The field work conducted in the Kendall Square CIC (**Figures 13–16**), complemented by interviews with CIC Relationship Managers [54, 55], proved the exceptional level of services provided to the companies hosted. The quality of the concierge, of the reception desks at each floor and of the complimentary meeting rooms and working spaces is outstanding. Flexible and high quality spaces are offered for a reasonable cost, since prices range from $425 to $1500/ person/month all included (stocked kitchens, conference rooms, Internet, printing & copying, phones, high-end furniture, operational & technical support and concierge). However, it is the style of management of the spaces that really makes the difference in conceptual terms. Each floor is equipped with a common kitchen, offering free food that varies on each floor in order to encourage people to move around the building. Cross-fertilisation of innovative ideas is considered the major asset offered to the hosted companies that are mixed across the floors and not clustered by sector. Collaboration complements cross-fertilisation, since companies looking for specialised services might find them within the CIC itself, resting on the assistance of the Relationship Managers, whose presence is ensured at all floors. It is the personal knowledge of the Relationship Managers with the individual company that orients the potential advice. The institutionalised role of a facilitator is a concrete example of coupling proximity with the opportunity of experiencing frequent and observable interactions. Further informal interviews with local stakeholders [45, 46, 56] confirmed the extraordinary role played by the CIC in building a favourable physical ecosystem, spurring innovation and supporting start-ups creation. Also, the CIC allowed some large companies, such as Google, to temporarily settle down in the area of Kendall Square prior to making the final decision of moving there with the entire headquarters.

Figure 14. Complimentary meeting rooms in the Kendall Square CIC building (author's picture).

Figure 15. Complimentary shared spaces (author's picture).

Figure 16. Venture Café (author's picture).

In conclusion, it is essential to complement S3 implementation with spatial interventions of the built environment that may facilitate the construction of a *physical ecosystem* supportive of innovation. These spaces include: shared spaces and private small businesses facilitating interaction, both informal and formal, both specialised and multi-disciplinary; public services and facilities that allow preserving uniqueness and inclusiveness. Furthermore, the spatial pattern has to be supportive of a walkable environment, offering effective transit and public transport facilities. The economic benefit for the private companies located in such areas is proved by the empirical findings from the case study, although not yet quantified.

Further research development includes the effort to quantify with monetary proxy the extra benefits above mentioned, incorporating in the assessment the public services and facilities in the area. This goal will be achieved by spatialising clusters first, then companies at the urban scale, then mapping the network of spaces that are supportive of social innovation and entrepreneurial discovery.

Acknowledgements

This chapter has been developed thanks to the field work made possible within the research project MAPS-LED, Multidisciplinary Approach for Planning Smart specialisation strategy for Local Economic Development, granted by the European Commission through the Horizon 2020 RISE program, project number: 645651. The author thanks all the interviewed experts, that generously provided her with suggestions and comments that were essential in developing the chapter and the research team based at the Senseable City Laboratory, MIT, Cambridge Boston, for the valuable insights gathered during the research presentation held in April 2016. However, the author is entirely responsible for any statement and any possible misunderstanding is her own responsibility. This chapter systematises and puts forward the preliminary findings of the author from the first year research activity on MAPS-LED, developed in the US and complemented by further research conducted in the UK and in Italy. This latter has been developed thanks to the teaching activities of the author both as PGR supervisor at SOBE and on the International Doctorate on Urban Regeneration and Economic Development. The author thanks the UK and Italian researchers, who contributed to build the UK and Italian exploratory case studies, in particular Clare Devaney, Vincenzo Crea, Arnauld Morisson and Giuseppa Romeo.

Author details

Claudia Trillo

Address all correspondence to: C.Trillo2@salford.ac.uk

School of the Built Environment, UPRISE University of Salford, The Crescent, Salford, Greater Manchester, UK

References

[1] European Commission. Communication from the Commission. Europe 2020. A strategy for smart, sustainable and inclusive growth. COM (2010) 2020

[2] Midtkandal I. and Sörvik J. What is Smart Specialisation?, in: Nordregio News Issue 5; 2012 Available from: http://www.nordregio.se [Accessed 2016-03-31]

[3] OECD. Innovation driven growth in regions. the role of smart specialisation. Paris: Organisation for Economic Growth and Development; 2013

[4] Foray D. Smart Specialisation: Opportunities and Challenges for Regional Innovation Policy. London: Routledge, 2015

[5] McCann, P. The Regional and Urban Policy of the European Union: Cohesion, Results-Orientation and Smart Specialisation. Edward Elgar Publishing Limited, Cheltenam, UK and Northampton, MA, USA; 2015

[6] Foray D., David P.A., Hall B. Smart specialisation: the concept, in: Knowledge for Growth Prospects for Science, Technology and Innovation. Selected Papers from Research Commissioner Janez Potočnik's Expert Group, European Commission. 2009. pp. 20–24: p.21

[7] European Commission. National/Regional Innovation Strategies for Smart Specialization (RIS3); 2014. Available from: http://ec.europa.eu/regional_policy/sources/docgener/informat/2014/smart_specialisation_En.pdf, pp. 3–4. [Accessed 2016-03-31]

[8] European Commission. The role of clusters in smart specialisation strategies. DG Research and Innovation; 2013. Available from: https://ec.europa.eu/research/evaluations/pdf/archive/other_reports_studies_and_documents/clusters_smart_spec2013.pdf, p.4 [Accessed 2016-03-31]

[9] http://s3platform.jrc.ec.europa.eu/ [Accessed 2016-03-31]

[10] Becattini G. Dal "settore" industriale al "distretto" industriale. Alcune considerazioni sull'unità d'indagine dell'economia industriale. Rivista di Economia e Politica Industriale , v. 1, 1979. pp. 7–21

[11] Becattini G. Sectors and/or districts: some remarks on the conceptual foundations of industrial economics, in: E. Goodman, J. Bamford (eds.), Small Firms and Industrial Districts in Italy. London: Routledge; 1989. pp. 123–135

[12] Marshall A. Principles of Economics, 8th ed. London: Macmillan; 1920

[13] Gravenotter M. Economic action and social structure: the problem of embeddedness.: American Journal of Sociology, 91, 1985. pp. 481–510

[14] Porter M. The economic performance of regions, Regional Studies 37, 6–7, 2003. pp. 549–578, p.562

[15] Porter, M. The Competitive Advantage of Nations. New York: Free Press; 1990

[16] http://www.clustermapping.us/ [Accessed 2016-03-31]

[17] Interview with Christian Ketels, Harvard Business School, 6 April 2016

[18] http://ec.europa.eu/growth/smes/cluster/observatory/cluster-mapping-services/cluster-mapping/mapping-tool/index_en.htm [Accessed 2016-03-31]

[19] http://www.clustercollaboration.eu/ [Accessed 2016-03-31]

[20] Haussmann R. and Rodrik D. Economic development as self-discovery. Journal of Development Economics, 72, 2, 2003, pp. 603–633

[21] Foray D. and Goenaga X. The Goals of Smart Specialisation. European Commission, Joint Research Centre; 2013. Available from: http://ftp.jrc.es/EURdoc/JRC82213.pdf [Accessed 2016-03-31]

[22] European Commission. Regional and Urban Policy, Guide to Social Innovation; 2013, p.6; p.59. Available from: http://s3platform.jrc.ec.europa.eu/documents/20182/84453/ [Accessed 2016-03-31]

[23] Porter, M. Clusters and the New Economics of Competition: Harvard Business Review, November–December Issue, 1998. Available from: https://hbr.org [Accessed 2016-03-31]

[24] Porter M. and Kramer M. R. Addressing a social issue with a business model, shared value creating shared value. Harvard Business Review, 2011. Available from: https://hbr.org/2011/01/the-big-idea-creating-shared-value [Accessed 2016-03-31]

[25] Barca F. An Agenda for a reformed cohesion policy. A place-based approach to meeting European Union challenges and expectations. Independent Report prepared at the request of Danuta Hübner, Commissioner for Regional Policy; 2009, p.5

[26] Foray D. Speech at the International Open Panel Discussion—MAPS-LED Research Project H2020. Reggio Calabria, Italy, 27–28 May 2015. Available from: https://www.youtube.com/watch?v=dG9rzNE-JmE [Accessed 2016-03-31]

[27] Barca F., McCann P. and Rodríguez-Pose A. The case for regional development intervention: place-based versus place-neutral approaches. Journal of Regional Science, 52,1,2012,134–152, p.147

[28] World Bank. World Development Report. Reshaping Economic Geography. Washington, DC: World Bank; 2009

[29] Sapir A., Aghion P., Bertola G., Hellwig M., Pisani-Ferry J., Rosati D., Vinals J. and Wallace H. An Agenda for a Growing Europe: The Sapir Report. Oxford: Oxford University Press; 2004

[30] OECD. Regions Matter: Economic Recovery, Innovation and Sustainable Growth. Paris: Organisation for Economic Growth and Development; 2009

[31] European Commission. European Spatial Development Perspective: Towards Balanced and Sustainable Development of the Territory of the EU. Luxembourg: Office for Official Publications of the European Communities; 1999

[32] Faludi A., Waterhout B. The Making of the European Spatial Development Perspective: No Masterplan. London and NY: Routledge; 2002

[33] Trillo C. Territorial dimension and spatial planning, in: Bevilacqua C. & Trillo C. Spatial planning and territorial development. Territorial dimension, development policies, planning. Napoli: Giannini Editore; 2012, pp. 59–98

[34] Faludi A. Speech at the International Open Panel Discussion—MAPS-LED Research Project H2020, Reggio Calabria, Italy, 27–28 May 2015; 2015. Available from: https://www.youtube.com/watch?v=dG9rzNE-JmE [Retrieved March 2016]

[35] Ekonomiaz. Special issue on Smart Specialisation Strategies. N°84, 2013

[36] http://www.cluds-7fp.unirc.it/perspectives_maps-led.php [Accessed 2016-03-31]

[37] Katz, B. and Bradley, J. The Metropolitan Revolution: How Cities and Metros Are Fixing Our Broken Politics and Fragile Economy. Washington, DC: Brookings Institution Press; 2013: p.55

[38] Informal exploratory interview with Bruce Katz, Jason Hachadorian and Scott Andes, Brookings Institute, Washington, D.C., 5 April 2016

[39] Novak M.A. Supercooperators. Altruism, Evolution and Why We Need Each Other to Succeed. NY and London: Free Press; 2011

[40] Rand D.G., Yoeli E. and Hoffman M. Harnessing reciprocity to promote cooperation and the provisioning of public goods. Policy Insights from the Behavioral and Brain Sciences, 1, 1, 2014, 263–269

[41] Informal exploratory interview with Erez Yoeli, Harvard, 7 April 2016

[42] CDD (Cambridge Community Development Department). Kendall Square Final Report; 2013. Available from: http://www.cambridgema.gov [Accessed 2016-03-31]

[43] Blanding M. The past and future of Kendall Square. A transformation in five acts, MIT Technology Review, 18 August 2015. Available from: https://www.technologyreview.com [Accessed 2016-03-31]

[44] Logan T. A new Kendall Square envisioned in $1.2b MIT plan, 28 July 2015. Available from: https://www.bostonglobe.com [Accessed 2016-03-31]

[45] Informal exploratory interview with Iram Farooq, Cambridge Community Development Department, 11 April 2016

[46] Informal exploratory interview with Jennifer Conway, MIT Investment Management Company, 14 April 2016

[47] http://www.cambridgema.gov/CDD/econdev/aboutcambridge/kendallsq, p.51 [Accessed 2016-03-31]

[48] Informal exploratory interview with Barry Bluestone, Dukakis Centre, Northeastern University Boston, 4 April 2016

[49] Florida R. The Rise of the Creative Class. And How It's Transforming Work, Leisure and Everyday Life. NY: Basic Books; 2002

[50] Timmerman L. Biogen Idec CEO on move back to Cambridge: "We're Working on It," 16 June 2011. Available from: http://www.xconomy.com [Accessed 2016-03-31]

[51] Schroeder B. How to build a biotech renaissance: MIT in Kendall Square. A look back at how Institute Professor Phillip Sharp, his startup Biogen, and MIT's biotech community helped revive Kendall Square. MIT News, 27 March 2014. Available from: http://news.mit.edu [Accessed 2016-03-31]

[52] MassTech Collaborative. Informal exploratory interview with Cem Oruc and Bethann S. Steiner, The Innovation Institute, 13 April 2016

[53] Cambridge Innovation Centre, http://cic.us [Accessed 2016-03-31]

[54] Informal exploratory interview with Sara Mruz, 15 April 2016

[55] Informal exploratory interview with Naomi Berlin, 15 April 2016

[56] Informal exploratory interview at the Kendall Square Association, 15 April 2016

3

Mapping the Land-Use Suitability for Urban Sprawl using Remote Sensing and GIS under Different Scenarios

Onur Şatir

Abstract

Urbanization is one of the important issues in fast developing countries, such as China, Turkey, Brazil, and South Africa. Therefore, sustainable urbanization strategies come into question while designing the cities. In this point, land-use suitability mapping for urban areas is of importance. Spatial information sciences, such as geographical information systems (GIS) and remote sensing are applied widely for mapping land-use suitability. In this study, Van City, which is the most crowded city in eastern Turkey, was evaluated by applying three different scenarios called ecological, economic, and sustainable. The multi-criteria evaluation technique was used in GIS environment in the mapping stage. Distance from roads, distance from urban boundary, hillshade, slope, elevation, land-use cover, and land-use ability factors were used as inputs in the analysis stage. The weights of each input factor were calculated according to urban change dynamics between 2002 and 2015. As a result of the study, the weighting approach using the natural change dynamics of Van City has a great potential to define objective weights. In addition, Van City was developed orderly on agricultural lands and grasslands, and it was not a sustainable development for the region because the main income is still agriculture and animal production, so a new strategy was designed in a sustainable scenario to prevent agriculture and grassland area loss in a mutual benefit between nature and human.

Keywords: multi-criteria evaluation, urban land-use suitability, fuzzy standardization, ideal point-based weighting, remote sensing, geographical information system, Van City, Turkey

1. Introduction

Human and nature are always in a paradox after the industrial revolution. Human requirements, such as food and accommodation, have increased in time because of raised human population and fast industrial developments. Particularly, fast developing countries, such as China, Turkey, Brazil, and South Africa, needed new lands for urban sprawl and industrial facilities recently. Landscape planning is important for decision makers in this point, because available lands for agriculture, urbanization, and industrial activities are limited and land-use planning and decision-making process are vital for a sustainable development.

In this content, land suitability analysis is one of the oldest forms of decision-making support systems in the field of landscape planning [1]. Nowadays, geographical information system (GIS) and remote sensing (RS) based techniques are used frequently in land-use planning to obtain land-use decisions and alternatives geographically. RS science is generally used to create layouts for spatial decision support systems (SDSS) in the analysis stage. RS provides time energy and cost savings, and huge areas can be identified using satellite images, radar and lidar technologies, and aerial photos from different platforms, such as unmanned air vehicles, towers, balloons, or planes [2]. RS can provide many data on the ecosystem physical, biological, and social dynamics to land-use planning studies. For example, topographical dynamics [3, 4], vegetation dynamics [5, 6], LUC dynamics [2, 7], soil physical, chemical, and biological dynamics [8, 9], and hydrological dynamics can be quantified using several modeling or classification techniques by RS.

RS outputs can be integrated with other raster or vector data in GIS interface to derive land-use decisions. When the decision support systems are integrated with GIS, it is called SDSS. SDSS is allowed stakeholder participation, iterative analysis, and integration with external spatial or non-spatial data sources [10]. Scientific literature has shown that SDSS is mostly used in multi-criteria evaluation (MCE) analyses. MCE is a multiple data assessment technique, and it is used not only in land-use evaluation studies but also in environmental risk probability mapping, such as landslide [11], forest fire risk [12, 13], and flood risk [14], in landscape planning science. GIS-based MCE techniques have three essential stages: criteria selection (factor definition), data standardization, and weighting [15].

"Criteria selection" is defined with respect to the study goal. All factors may be called as inputs. In this extent, this stage is important to get an accurate result. These data sets must identify the essentials of the research. For example, in an agricultural land-use suitability mapping for wheat crop, inputs are related on wheat growth environment, such as climatic variables (temperature, precipitation, solar radiation, humidity), physical variables (elevation, aspect, slope, hillshade, soil depth, soil texture), chemical variables (soil pH, soil nutrients, soil salinity), accessibility (road network, settlement location), and bio-environmental variables (chlorophyll content, crop yield). Additionally, environmental risk factors, protection areas (natural and historical sites), and restriction areas (roads, security regions, built-up areas) must be defined spatially.

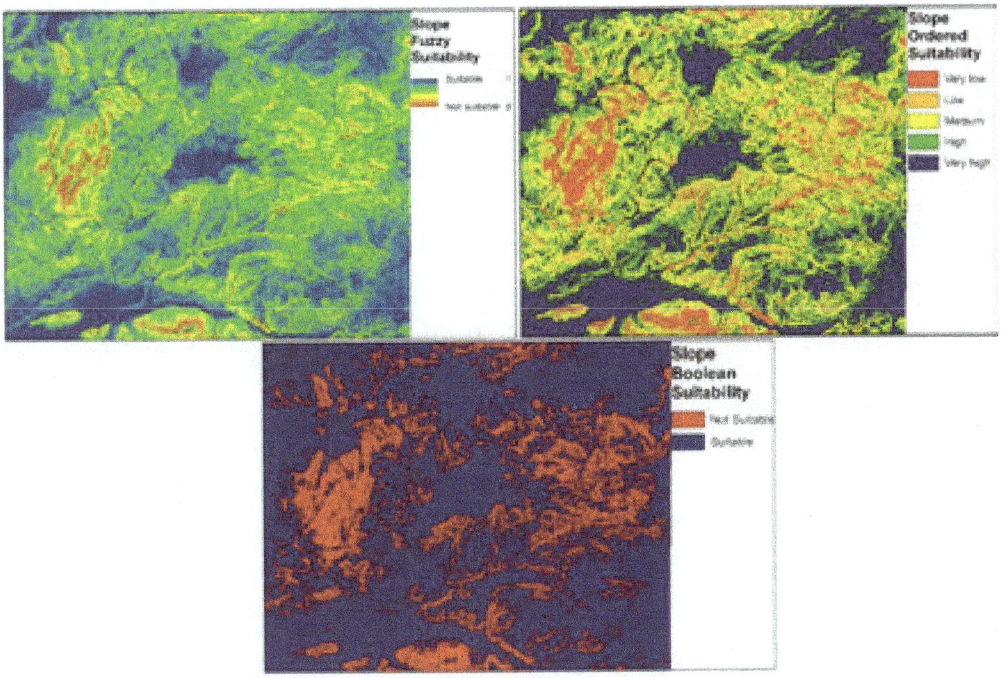

Figure 1. Sample slope standardizations using various techniques for urban sprawl suitability.

"Standardization" is necessary because of input data unit and range differences. Inputs will multiply with detected weights and standardization is a vital part for input data comparison. However, some non-parametric multiple data assessment techniques, such as artificial neural network (ANN) or regression tree, do not need standardization process because of internal weighting abilities [13]. There are many standardization techniques, such as fuzzy standardization, linear ordered standardization, and Boolean approach. Introducing these techniques with a sample is better for understanding. Imagine a slope map of a land in urban built-up suitability study. The slope unit is degree and data range is between 0 and 90°. In the fuzzy approach, flat fields defined the most suitable (1) and steep fields defined the opposite (0). Other fields are valued between 0 and 1 according to the suitability values, such as 0.1, 0.2, 0.25, etc. So that there will not be any absolute transition zone in the standardized map, all transitions will be soft surface. This technique is appropriate to avoid the missing data effect in transition zone. Ordered standardization needs categorical data. In this technique, slope data must be reclassified based on slope degrees, such as 0-5, 5-15, 15-25, 25-35, etc. Each slope range is reclassified as 1, 2, and 3 based on slope suitability for urban built-up. In this method, there is no transition zone and the boundaries of the categories are hard. In Boolean approach, all inputs are standardized as only 1 and 0 or suitable or not. There is a threshold value to detect the suitable land for each factor. For example, if slope degree less than 15 is suitable, other areas are not suitable for urban built-up. Slope data standardization using these techniques is shown in **Figure 1**.

"Weighting" is the hardest part of an MCE analysis. Here, expert-based weighting, literature-based weighting, and ideal point-based linear weighting are discussed. In addition, there are some non-linear ideal data-dependent techniques, such as ANN, logistic regression, ant colony algorithm, and regression tree.

Expert-based technique is preferred by those who have not any ideal point data, because, in this technique, all factors are evaluated by the experts, applying a survey that includes questions about the priority of the factors. Analytical hierarchical process (AHP) can be used because of pairwise comparison abilities to detect the weights of factors after the expert's decisions. Pairwise comparison matrix is defined as weights using binary priority definition ability [16, 17]. Although expert-based weighting is easy, all experts may be given various answers to the same question. Thus, the subjectivity of this method is high. However, this method is still used widely [18].

Literature-based weighting is another approach to define factor weights. It is completely based on similar studies in the literature and factor weights are adapted to the new research from previous studies. This method is easy for weight definition; however, regional environmental and social differences are ignored in this approach, so the reliability of the technique is a question.

Ideal data-based weighting is the most reliable approach in MCE studies. The problem in this technique is that what is the ideal data. For example, in a crop-based agricultural land-use suitability detection study, crop productivity can be used to define ideal crop suitability areas and all factors can be weighted relating to high productive areas to find potential suitable areas. Crop productivity data can be obtained from farmer surveys or using RS [13]. There is another technique using ideal point for weighting called the weight of evidence (WOE). This approach is also effective to define factor weights in land-use suitability or environmental hazard probability studies [19, 20].

This chapter discusses land-use suitability for urban growth using ideal point-based technique. Ideal urban sprawl is defined using urban growth dynamics from past to current. In this extent, Van City, which is located in eastern Turkey, was modeled by applying three different scenarios: economic, ecological, and sustainable. Distance from road and central city area, elevation, slope, hillshade, land-use ability (LUA), and land-use cover (LUC) were used as factors in the study. Weights were defined according to urban change and were used in all scenarios. On the contrary, restriction areas and fuzzy suitability degrees of LUC and LUA data were modified separately for the scenarios.

2. Short background: GIS-based land-use suitability studies

While factor and objectives are spatial, a GIS interface is necessary to analyze geographic data, and it requires a combination of multi-criteria methods with a GIS interface [21, 22]. In the literature, there are many land-use decision studies using GIS and other spatial information science, such as RS. Malczewski [22] completed a detailed study on the survey of the GIS-based MCE techniques. Results showed that this method is used widely (top three subjects) and orderly in environment and ecology, transportation, and urban-rural planning studies. In addition, these techniques are used in waste management, agriculture and forestry, facility area suitability, rangeland management, recreation and tourism, natural hazard studies, hydrological studies, and real estate-housing studies. Some techniques, data, and criteria used for land-use suitability analyses are found in **Table 1**.

Source	Technique	Factors	Suitability field	Year
[31]	GIS-based ANN	Soil depth, moisture, fertility, texture, salinity, aeration, temperature, accessibility	Agriculture	1994
[32]	Boolean overlay method	Slope, soil mechanics, flood-erosion hazard, water level and drainage, toxicities, rooting conditions	Crop growth suitability	2002
[33]	AHP	Soil depth, LUA, erosion hazard, slope, elevation, distance from water sources, distance from road, limiting soil factors	Land use	2009
[34]	GIS-AHP	Population density, available land, land values	Public parks	2011
[35]	GIS-AHP expert-based weighting	LUC, DEM, slope, tourist map, attraction places, road map, population, protected areas, wildlife areas	Tourism and recreation suitability	2011
[36]	GIS-MCE-AHP expert-based weighting	Accessibility, slope, interactions with other facilities, population density	Educational land-use	2011
[15]	RS integrated fuzzy, GIS-AHP, productivity-based weighting	Temperature, precipitation, soil texture, soil salinity (EC), soil depth, soil porosity, GDD, crop productivity	Crop growth suitability	2013
[18]	AHP, expert-based weighting	Soil groups, soil depth, erosion, slope, aspect, elevation, soil parameters, LUA	Agriculture	2013

Source	Technique	Factors	Suitability field	Year
[37]	RS integrated GIS, expert-based weighting using AHP	Slope, soil depth, texture, moisture, organic C, pH, EC, primary nutrients (N, P, K)	Agriculture	2015
[30]	GIS-MCE-fuzzy AHP	Residential, extractive industry, marine industry, recreation subfactors	Urban land suitability	2015
[38]	RS integrated GIS-based MCE, AHP	NDVI, LUC, climate data set, DEM, economic and social data	Eco-city evaluation	2015
[39]	Fuzzy logic, AHP-GIS-WLC	Distance from river, LUC, urban areas, crop pattern, distance from residential areas, roads, proximity to interested areas	Waste water disposal area suitability	2016

EC, electrical conductivity; GDD, growing degree days; NDVI, normalized difference vegetation index.

Table 1. Sample techniques and factors used in land-use suitability studies.

According to **Table 1**, even if the suitability target is same, there might be small differences between factors. Some of the important essential factors are the same in agricultural suitability studies, such as soil depth, slope, and soil texture. However, there are different factors: erosion rate, soil nutrients, and crop productivity. Data accessibility, database availability, and method can be affected factors despite the same purpose. Also, these factors may be changed with respect to regional differences. If there is flood risk in an area, we have to assess flood risk rate in land-use suitability analyses.

3. Study area

Van City is located in eastern Turkey, and it is the most crowded city of the Eastern Anatolian Region of Turkey (**Figure 2**). Van Province's population was 1,096,397 in 2015, and almost 472,000 of them lived in Van central city area. Population has increased almost 5.5% during the last 5 years [23]. The main incomes in the region are animal production, agriculture, border trading, and tourism. Mainly continental climate is dominant, but alpine and sub-alpine climate effects are observed in high regions. Van central city's area climate is warmer than in those around it because of Van Lake. This lake is the biggest lake in Turkey and its water contains $CaCO_3$. The central city area of Van was selected as the study area due to the fast development in last 10 years. The city was damaged by two big earthquakes (7.2 and 6.1) in 2011. After these tragic events, urban area started to sprawl to the far regions from the central city area.

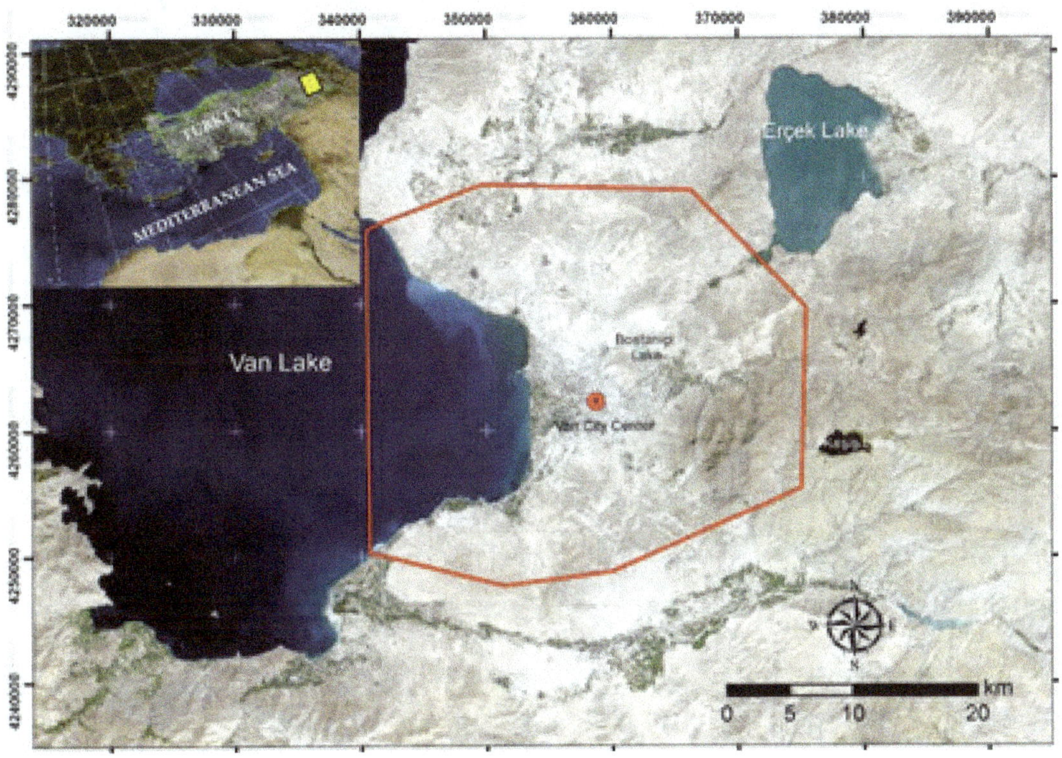

Figure 2. Location of the study area and boundary of the case area (UTM WGS 84 Projection Zone 38N).

4. Data characteristics

The research data set contains (i) satellite images and LUC data, (ii) digital elevation model (DEM) data, (iii) LUA data, (iv) distance from road and urban built-up area data, and (v) restrictive areas or constrained areas (**Figure 3**).

4.1. Satellite images

Landsat 5 TM and Landsat 8 OLI data sets were used to detect urban area change between 2002 and 2015 in Van central city area. Landsat imageries have a great potential for monitoring the land-use/cover change because of large time series database and available spatial (30 m) and spectral (VIS, NIR, SWIR, and TIR) resolutions [24]. Two Landsat images were used for LUC and change detection. An earlier image was taken on August 2002, and the later one was recorded on August 2015.

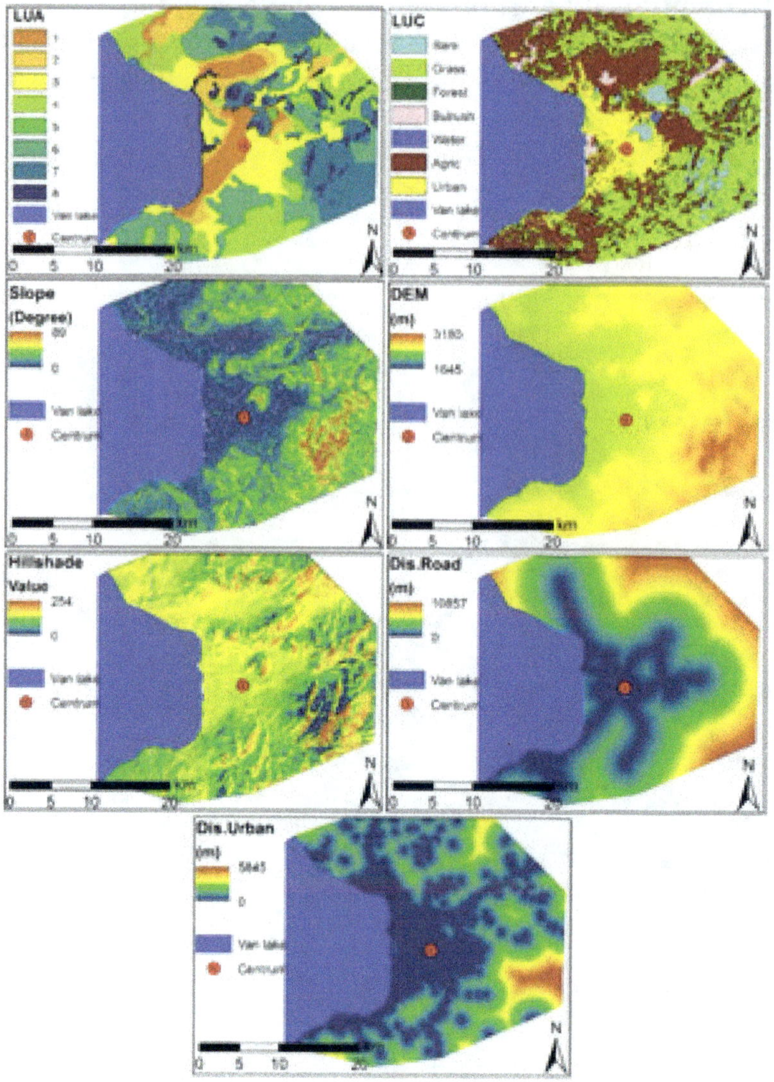

Figure 3. Input data set (criteria or factors) used in the urban sprawl suitability of Van City.

4.2. DEM data

Elevation data were obtained from ASTER Global DEM data set in 30 m spatial resolution. Slope and hillshade images were produced from DEM data in GIS interface. Additionally, DEM data were used as ancillary data in LUC classification stage to detect the LUC change based on elevation. Also, such data sets may be improved for the accuracy of the LUC classification [2].

4.3. LUA data

The LUA of Turkey was mapped by Soil Survey Staff of Turkey. This map shows capable lands and capability degrees in eight categories (1 refers to very capable and 8 refers to not capable) based on slope, soil depth, soil type, geological type, etc. Highly available lands can be used for agriculture, urban built-up, and industrial facilities. However, land-use suitability is variable according to the scenarios. For example, agricultural areas are important because available lands are limited so, in ecological or sustainable scenarios, highly available lands must be protected for agriculture.

4.4. Distance images

Distance from roads and urban built-up areas are produced from road maps and urban area maps, which are produced from classified LUC map. Road and settlement distance is significant to evaluate the infrastructure and superstructure availability of a land for urban sprawl.

4.5. Restrictive areas

Urban cannot sprawl to regions, such as water bodies, existing built-up areas, security areas, and historical protection areas. Also, restrictive areas are modified according to the scenarios. Some wetlands are closed for urban sprawl in the ecological scenario.

5. Methodology

The study was performed in two substages: (i) LUC classification, change detection, and accuracy assessment and (ii) MCE process in GIS environment (**Figure 4**).

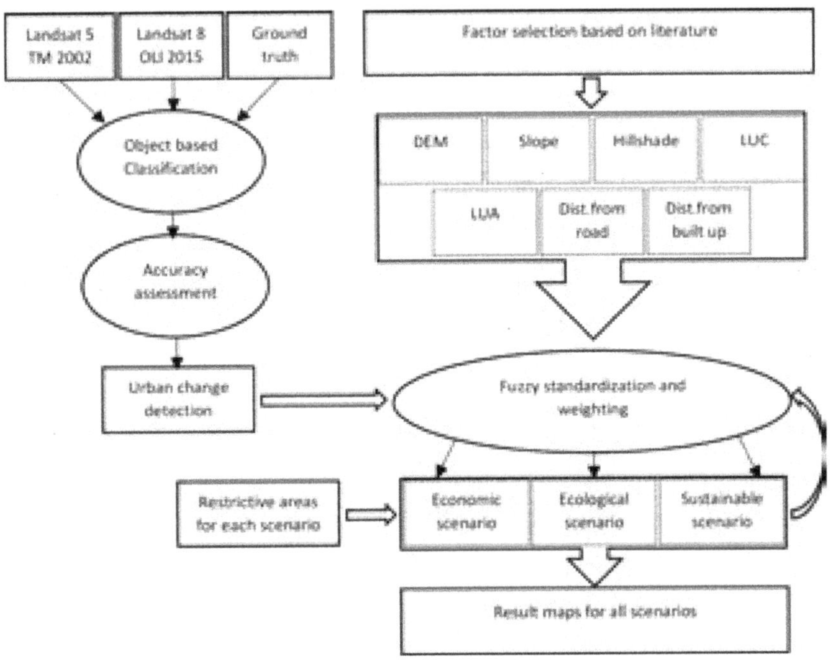

Figure 4. Summary of the methodology.

5.1. Object-based classification

Many complex land covers exhibit similar spectral characteristics, making separation in feature space by simple per-pixel classifiers difficult, leading to inaccurate classification. Therefore, an object-based classification is a potential solution for the classification of such regions. The specific benefits are an increase in accuracy, a decrease in classification time, and that it helps to eliminate within-field spectral mixing [25]. Basically, there are three steps in object-based classification: segmentation, classification, and per field integration. An image is divided into segments dependent on pixel spectral similarities, structure of the image, and surface texture characteristics. This progress is up to variables such as scaling factor, smoothness versus compactness, and shape factors. Each segment contained a group of pixels, and scaling factor is defined as the minimum pixel counts that have similar spectral characteristics in a segment. Compactness and smoothness are important for creating pixel groups. Shape factor deals with the boundary of a segment. Scale factor is variable according to the study scale and ideal scale can be found trying different scale factors [2].

5.2. Accuracy assessment of LUC classifications

The accuracy of the LUC classifications is tested by applying error matrix and κ statistic. The error matrix approach is the most widely used in accuracy assessment [26]. After the generation of an error matrix, other important accuracy assessment elements, such as overall accuracy, user accuracy, producer accuracy, and κ coefficient, can be derived. κ is the difference between the observed accuracy and the chance agreement divided by 1 minus that chance agreement [27]. Ground truth data collected from field trip and old topographical and forest maps are used for the accuracy assessment of 2002 and 2015 LUC.

5.3. GIS-based MCE process

The general information on this process is discussed in Section 1. However, criteria selection, standardization, weighting, and mapping progress and methods are introduced in urban sprawl suitability extent in this study. Factors are defined based on similar literature, data accessibility, and regional variations. Seven factors are defined: elevation, slope, hillshade, LUC, LUA, distance from roads, and distance from built-up areas.

Weighted linear combination (WLC) is one of the most used MCE techniques to define suitability degrees for continuous factors. With WLC, factors are combined by applying a weight to each followed by a summation of the results to produce a suitability map [28]:

$$S = \sum WiXi, \tag{1}$$

where S is suitability, Wi is the weight of factor i, and Xi is the criteria score of factor i.

5.3.1. Fuzzy approach and weighting

Nothing is exact in nature. Fuzzy mentality is based on this simple rule. In a landscape, boundaries are flexible and natural characteristics of a land do not change suddenly, such as slope and soil texture. A fuzzy set can define a factor with transition zones according to the suitability degree of the factor [29].

Factors are of different data ranges and units. Weighting a factor linearly needs a standard input data set to define the priority degree of the factors. Standardization process is applied to the factors separately. However, ideal and non-ideal values of the each factor have to be defined primarily. In this study, ideal regions for urban sprawl are defined based on urban development between 2002 and 2015. Changed regions are accepted ideal areas for urban sprawl. The ideal and non-ideal values of each factor are defined considering the existing urban growth. It does not mean that the existing urban growth is ideal for sustainable development, but, finally, Urban was developed to these regions and the spatial characteristics of the changed areas are good indicators for the future urban sprawl. All factors are reclassified based on ideal intervals, which are defined according to data frequencies, and ideal data ranges are found for fuzzy standardization.

The weights of each factor are obtained using ideal data diversity in the categories of the factors. For example, slope is reclassified to five intervals: very high, high, medium, low, and very low. If urban change is located in all areas homogeneously, it means that the slope is not important for urban sprawl; if urban change is monitored heterogeneously, it means that the slope is important because only one or two slope ranges are suitable for urban growth. The standard deviation (SD) of the each factor diversity is shown as data homogeneity with a standard result. If the SD value is high, weight should be high. The weights of each factor are defined according to the SD values.

5.3.2. Scenario development

Three scenarios were evaluated in the study: economic, ecologic and sustainable. In the economic scenario, only water bodies, existing city areas, security areas, and historical protection areas were analyzed and defined as restrictive regions. There was not any limitation on LUC or LUA. Therefore, the trend of the last 13 years was considered to define urban suitability in the economic scenario. Ecologically important areas, such as wetlands, were ignored.

In the ecologic scenario, important areas, such as wetlands (bulrush) and highly available lands for agriculture and grasslands, were protected strictly from urban sprawl.

In the sustainable scenario, urbanization demands were considered using urban change in time. According to urban change results, agricultural and wetland areas were protected partly in addition to other restrictive areas, so that a balanced urban spread was provided for human life and nature in the future.

6. Results

In this study, urban change in time has the key role for standardization and weighting of the factors. Therefore, RS science is into play in this point to create LUC layouts and define urban change. Then, a GIS-based MCE process is applied according to the changed areas.

6.1. Urban change detection by RS

Landsat imageries taken on August 2002 and 2015 were classified using object-based classification approach. A post-classification change detection technique was applied and atmospheric or radiometric correction is not necessary in this technique. Also, Landsat imageries have already been corrected geometrically. Segmentation parameters were modified after a small experimental application. Seven LUCs were classified, except Van lake area: agriculture, grassland, bareground, settlement, bulrush, woodland, and inland water.

August 2002 and 2015 overall κ classification coefficiencies were obtained as 0.89 and 0.92, respectively. The urban built-up area was 2066 ha in 2002 and increased to 7694 ha in 2015. Then, 3841 ha agriculture, 1265 ha grassland, 328 ha bareground, 89 ha woodland, and 18 ha bulrush areas were transformed to the urban. Particularly, after the 2011 earthquakes, urban growth system was changed and new residential areas were established on the far regions from the city center. This was affected urban built-up area change fast (**Figure 5**).

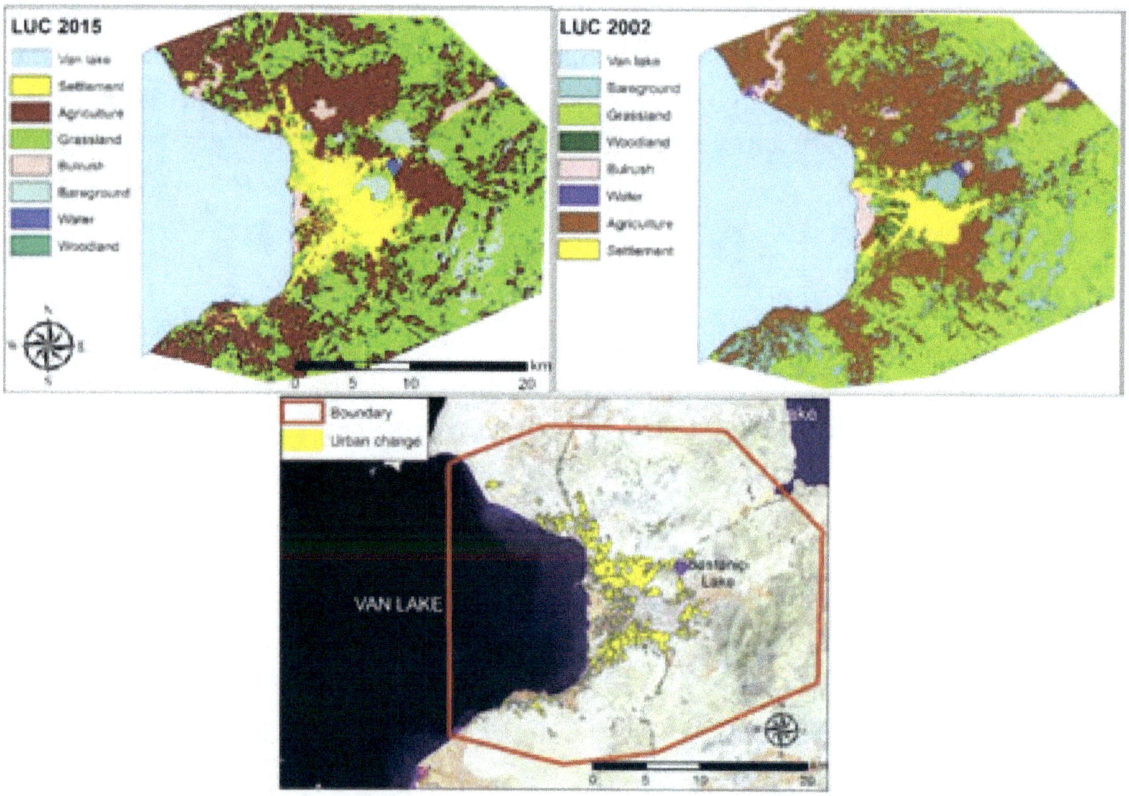

Figure 5. LUC of Van central city area (2015–2002) and changes in urban built-up areas.

6.2. Fuzzy standardization and factor weighting

Factors were reclassified based on natural breaks in ideal intervals. LUC and LUA data sets were already categorical. However, other inputs were continuous data format. Elevation data were reclassified into eight categories with 100 m intervals. Slope was divided into five categories with 15° intervals as very low (flat), low, medium, high, and very high. Hillshade was reclassified into five categories based on sun effect as very high, high, medium, low, and very low. Distance from roads and distance from urban built-up areas were classified as 10 intervals for each 1000 m. Urban area change was recorded as 5628 ha and all changed areas were separated according to urban change areal diversity to each category for each factor (**Table 2**).

Categories	1	2	3	4	5	6	7	8	9	10
Elevation*	4169	1188	161	76	26	19	13	0	—	—
Slope*	5631	20	0	0	0	—	—	—	—	—
Hillshade*	1	9	496	5109	37	—	—	—	—	—
Distance from road*	4293	863	177	124	6	39	91	14	17	4
Distance from built-up areas*	4367	532	257	168	137	62	39	33	28	5
LUA**	1783	973	1675	258	21	523	257	136	—	—
LUC***	3841	1265	328	89	74	—	—	—	—	—

*(1) Refers lowest value range of the factor.
**(1) Refers the highest value of the factor.
***(1) Agriculture, (2) grassland, (3) bareground, (4) woodland, and (5) bulrush.

Table 2. Areal diversity of urban change areas to each category inside the factors.

All factors must be evaluated alone based on the suitability degree for fuzzy standardization. In this frame, following fuzzy standardization functions were applied to the factors and all inputs were standardized between 0 and 1 (**Table 3**; **Figure 6**).

Categories	Technique	Function	Explanation
Elevation	Monotonically decreasing	Almost linear	When elevation is increased, urban suitability is decreased
Slope	Monotonically decreasing	User defined	When slope degree between 0 and 15 suitability is decreasing polynomial, urban growth is not possible after the 35th degree
Hillshade	Optimal value	Gaussian or symmetric in narrow range	Sun effect should be high but not very high or very low

Categories	Technique	Function	Explanation
Distance from road	Monotonically decreasing	Almost linear	If an area far from the road network, suitability of urban sprawl is decreased
Distance from built-up areas	Monotonically decreasing	Almost linear	If an area far from the road network, suitability of urban sprawl is decreased
LUA LUC	Optimal values	User defined	Suitability degree of each category was defined based on urban growth distribution from past to current

Table 3. Fuzzy standardization rules of the factor for urban built-up suitability.

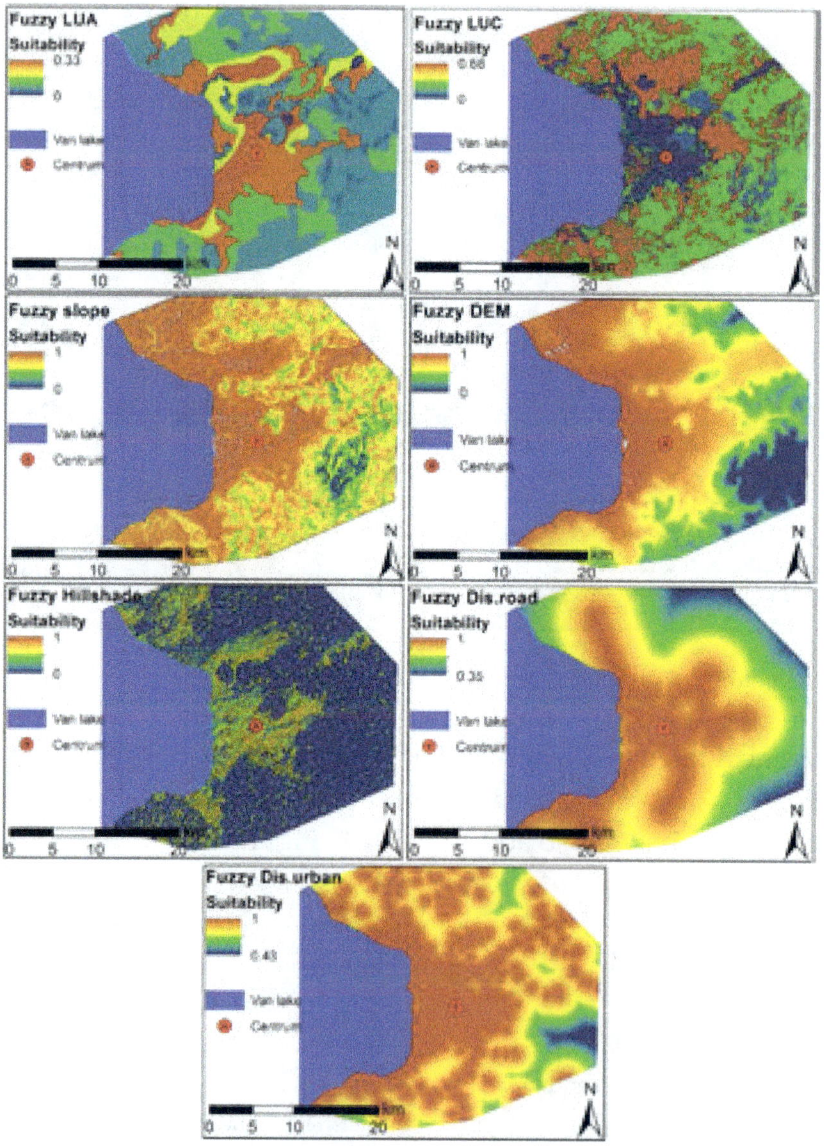

Figure 6. Standardized factors by fuzzy approach for urban sprawl suitability.

Weighting the factors was performed according to the SD of urban change in each factor. Areal diversity of urban change was already obtained for each category of factors. SDs were used to see the heterogeneity of urban change diversity in the factors. The SD values and weights are shown in **Table 4**.

Categories	SD	Weights
Elevation	1456	0.131
Slope	2516	0.227
Hillshade	2234	0.2
Distance from road	1336	0.12
Distance from built-up areas	1346	0.121
LUA	699	0.063
LUC	1498	0.135

Table 4. Weights of factors.

As a result of the weights, slope and hillshade were the most significant factors for urban sprawl. The lowest effect was recorded in LUA because the urbanization aspects from past to current were not the care of LUA for urban sprawl. However, these results showed that urban sprawl in time was not sustainable if this situation continued, because city growth on the agricultural areas that are limited in the region and there is no too much option for income, except agriculture and tourism in the region.

6.3. Scenario applications

Scenarios are described in Section 5.3.2. In the application stage, there are several approaches to apply a scenario. Some of these approaches were applied modifying the weights of each factor for each scenario. However, in this study, all factor weights have been already obtained based on urban change, and physical variables, such as hillshade, elevation, slope, and distance factors, are not changed according to the our scenarios. Therefore, the weights of these data were used the same in all scenarios. However, LUC and LUA suitability can be changed according to the economic, ecological, and sustainable scenarios.

In the economic scenario, the fuzzy standardization of the LUC and LUA was defined without any limitation. For example, agricultural areas were transformed to urban areas in 13 years dominantly. In the economic scenario, this situation was continued; in the ecological scenario, some of the agricultural areas were protected, which were located in the first, second, and third zones of the LUA. Also, in the ecological scenario, wetland, coastal line, and nearby and natural grassland usage was limited for urban sprawl.

In the sustainable scenario, only the first zone of the LUA was protected because, if first three zones were protected, there would not be enough urban sprawl area in the future and this situation would not be sustainable.

6.3.1. Urban sprawl suitability in the economic scenario

Restrictive areas were defined as historical protection areas, current built-up areas, water surfaces (lake and rivers), and security zones (military security areas). There was no limitation on LUC and LUA or wetland usage in the area. In this scenario, urban change was presumed continuous in the future like in the last 13 years.

Suitability degrees were classified into five categories and constrained areas. In the economic scenario, particularly, medium and high suitability regions covered 35% and 32%, respectively. Restrictive areas only covered 12% of the total area (**Figure 7**).

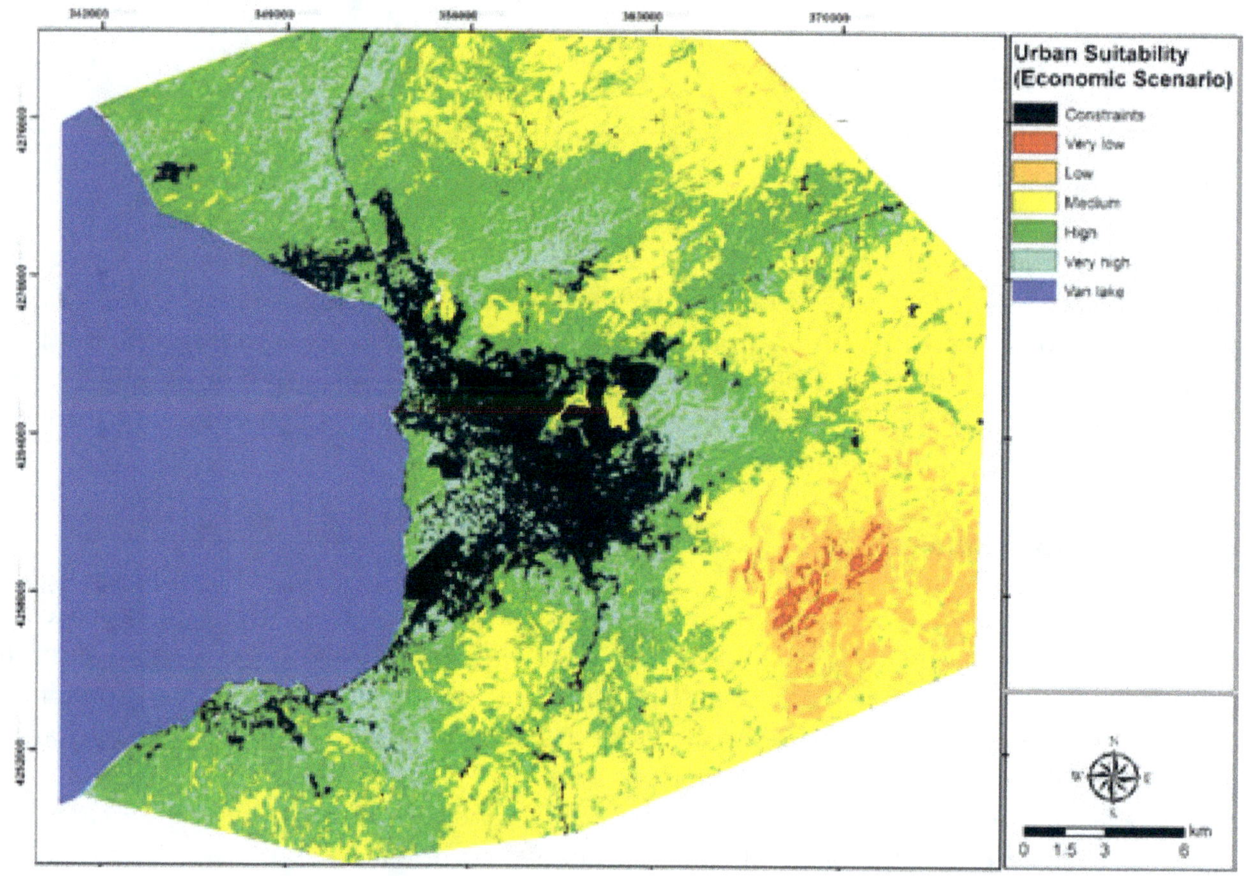

Figure 7. Distribution of the urban sprawl suitability under the economic scenario.

6.3.2. Urban sprawl suitability in the ecological scenario

This scenario was protected in all important natural lands and high capable lands for agricultural usage strictly. In this extent, the LUC map of 2015 and LUA maps were re-standardized and the lowest suitability degree was assigned to agriculture after the bulrush (wetlands) areas in the LUC map. In the LUA map, the first, second, and third zones were extracted from the result map, because these lands are available for agriculture.

Medium suitable lands and constrained areas covered 41% and 36%, respectively. High suitable lands were located in the north and east sides of the city mainly (**Figure 8**).

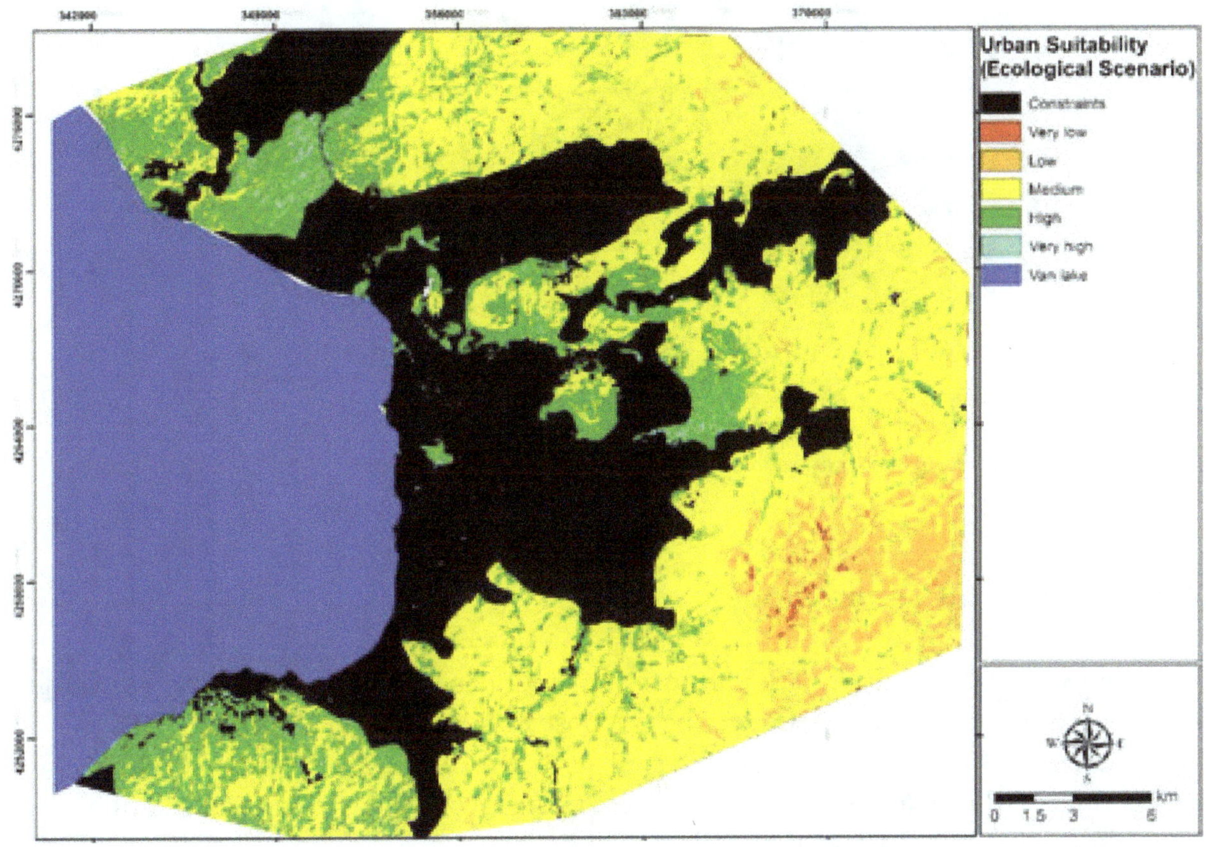

Figure 8. Distribution of the urban sprawl suitability under the ecological scenario.

6.3.3. Urban sprawl suitability in the sustainable scenario

This scenario is a mixture of the economic and ecological scenarios. Restrictive areas were less than the ecological scenario, because only the ecological scenario is not applicable and sustainable for a fast developing city like Van City. However, natural lands and coastal regions that have a good recreational ability were ignored in the economic scenario. Additionally, most of the agricultural areas were suitable for urban growth under the economic point of view because of construction expenses (less filing process and infrastructure necessity). Therefore, we have to both protect important areas for high-quality life in the future and answer the future urban built-up demand. Therefore, the sustainable scenario was considered in both variables.

Medium and high suitable areas covered 43% and 28% of the total area, respectively, in the sustainable scenario. Restrictive areas covered 19%. Agricultural areas were protected partly according to the suitability degree in the LUA map. Only the first zone was extracted from the result map, and the fuzzy suitability map of the LUA was defined based on LUA stages orderly in a negative way. Urban sprawl was planned on less productive agricultural areas without ignoring the urban sprawl on agricultural areas (**Figure 9**).

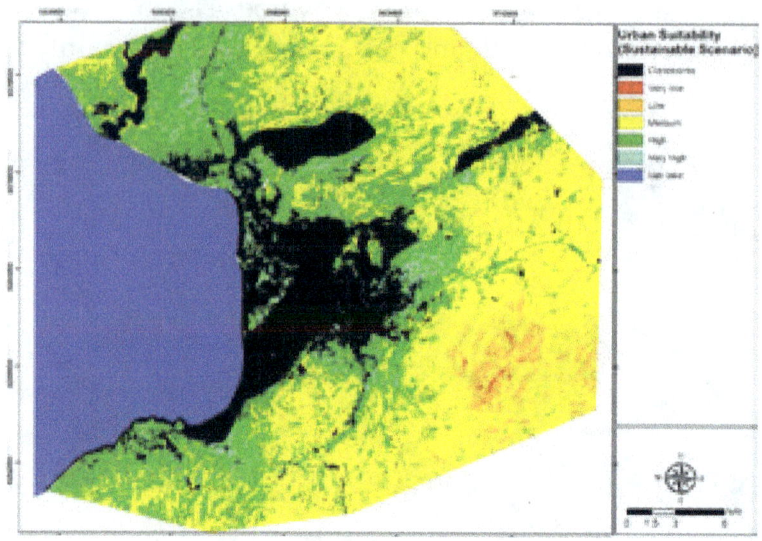

Figure 9. Distribution of the urban sprawl suitability under the sustainable scenario.

7. Discussion and conclusions

Urban sprawl suitability under the various scenarios was applied to Van City, which developed fast in the last decade in Turkey. The purpose of the study was not only to focus on the Van City growth options but also to present how we can apply various scenarios in GIS interface for a land-use decision-making process using an objective weighting technique based on an indicator data set such as urban change in time.

Suitability degree	Economic scenario		Ecological scenario		Sustainable scenario	
	Area (ha)	%	Area (ha)	%	Area (ha)	%
Constrained	8910	12	25,814	36	13,597	19
Very low	425	1	125	<1	62	<1
Low	4220	6	4125	6	2617	4
Medium	25,032	35	29,434	41	31,356	43
High	23,047	32	11,243	16	19,860	28
Very high	10,505	15	1406	2	4613	6

Table 5. Areal diversity of the urban suitability categories of scenarios.

Three different scenarios were cross-checked as economic, ecologic, and sustainable using seven factors, which have significant effects on urban sprawl. These factors were diversified according to the regional differences or data accessibilities for urban growth suitability [30]. As a result of the study, restrictive areas covered too much places in the ecological scenario. The highest suitability areal cover was recorded in the economic scenario. On the contrary, in the sustainable scenario, development areas were defined ideally for the future urban sprawl of Van City. Van City changed almost 5700 ha in 13 years, and it needs a minimum of 5000 ha area for the next 10 years according to the population increase. In the sustainable scenario, very high suitable lands almost satisfied this demand. Also, high suitable areas were enough for the long future development (**Table 5**).

In general, very high and high suitable areas were located in north, east, and south parts of the city. In the city development plan, a belt highway is under construction from the north to south parts of the city shaped as a half-circle. In this extent, heavy industrial places are located at the north side of the city and it will improve to near the belt highway in the northern side. These regions are available for industrial development. The south part of the city started to develop after the earthquakes in 2011. According to the ground surveys (ground stability map could not be used in this study because of data accessibility problem), this is available for urban sprawl without taking any additional construction preventions for built-up area growth. The east part of the city developed so fast after the 2000s. Particularly, immigrants from the rural areas and other cities settled in the east part of the city. Unfortunately, the east part is a good sample for unplanned urbanization, so this part will probably be within the scope of urban transformation in future. Therefore, urban sprawl in this region will be less than expected.

In conclusion, GIS-based MCE techniques are efficient to evaluate the land-use options because of geographic interface and scenario application abilities. Also, this visual ability can be integrated with public participation in weighting or decision-making stages. In this study, RS data were used to define fuzzy rules and weights of factors using a categorical diversity of changed areas in the factors. This approach was provided objective weighting by an ideal data set. This ideal data set was not an ideal city development, but finally the city developed on these areas in time and priority of the factors can be organized according to past urban development. For the ideal urban growth, scenarios were tested. Sustainable urban sprawl areas were detected comparing the ecological and economic scenarios and urbanization demand in the future. As a result of the decisions, a sustainable scenario was developed and applied to Van City considering sustainable life quality, urbanization demands, and nature protection.

Author details

Onur Şatir*

Address all correspondence to: osatir@yyu.edu.tr

Department of Landscape Architecture, Faculty of Agriculture, Yuzuncu Yil University, Van, Turkey

References

[1] Malczewski J: GIS-based land-use suitability analysis: A critical overview. Progress in Planning. 2004; 62(1): 3–65.

[2] Şatır O, Berberoğlu S: Land use/cover classification techniques using optical remotely sensed data in landscape planning. In Özyavuz M. (Ed.) Landscape Planning. Rijeka: InTech; 2012. pp. 21–54. DOI: 10.5772/31351.

[3] Guarnieri AM: SAR interferometry and statistical topography. IEEE Transactions on Geoscience and Remote Sensing. 2002; 40(12): 2567–2581.

[4] Zhanyu W, Arrowsmith JR, Honglin H: Evaluating fluvial terrace riser degradation using LIDAR-derived topography: An example from the northern Tian Shan, China. Journal of Asian Earth Sciences. 2015; DOI: 10.1016/j.jseaes.2015.02.016.

[5] Berberoglu S, Satir O, Atkinson PM: Mapping percentage tree cover from Envisat MERIS data using linear and non-linear techniques. International Journal of Remote Sensing. 2009; 30(18): 4747–4766.

[6] Donmez C, Berberoglu S, Curran P: Modelling the current and future spatial distribution of NPP in a Mediterranean watershed. International Journal of Applied Earth Observation and Geoinformation. 2011; 13: 336–345.

[7] Akin A, Sunar F, Berberoğlu S: Urban change analysis and future growth of Istanbul. Environmental Monitoring and Assessment. 2015; 187: 506.

[8] Satir O, Berberoglu S, Kapur S, Nagano T, Akça E, Erdogan MA, Donmez C, Satir NY, Tanaka K: Soil salinity mapping using Chris-Proba hyperspectral data. In Proceedings of ESA Hyperspectral Workshop 2010; 17–19 March 2010; Frascati, SP-683.

[9] Lakshmi V, James J, Soundaria S, Vishalini T, Pandian KP: A comparison of soil texture distribution and soil moisture mapping of Chennai coast using Landsat ETM+ and IKONOS data. Aquatic Procedia. 2015; 4: 1452–1460.

[10] Kim HY, Choi Y, Kim H, Oh SH: Planning for the sustainable? Land use suitability and social and ecological factors for locating a new hazardous facility. KSCE Journal of Civil Engineering. 2016; 20(1): 359–366.

[11] Chalkias C, Ferentinou M, Polykretis C: GIS-based landslide susceptibility mapping on the Peloponnese Peninsula, Greece. Geosciences. 2014; 4: 176–190.

[12] Jaiswal RK, Mukherjee S, Raju KD, Saxena R: Forest fire risk zone mapping from satellite imagery and GIS. International Journal of Applied Earth Observation Geoinformation. 2002; 4: 1–10.

[13] Satir O, Berberoglu S, Donmez C: Mapping regional forest fire probability using artificial neural network model in a Mediterranean forest ecosystem. Geomatics Natural Hazards and Risk. 2015; DOI: 10.1080/19475705.2015.1084541.

[14] Papaioannou G, Vasiliades L, Loukas A: Multi-criteria analysis framework for potential flood prone areas mapping. Water Resources Management 2014; 29(2): 399–418.

[15] Şatır O: Determining the agricultural land use suitability using remote sensing and geographical information system in Lower Seyhan Plane (Ph.D. thesis). Cukurova University Natural and Applied Sciences Institute; 2013.

[16] Saaty T. The Analytical Hierarchy Process. New York: John Wiley; 1980.

[17] Saaty T: Relative measurement and its generalization in decision making: why pairwise comparisons are central in mathematics for the measurement of intangible factors e the analytic hierarchy/network process. Review of the Royal Spanish Academy of Sciences Series A Mathematics. 2008; 102(2): 251–318.

[18] Akıncı H, Özalp AY, Turgut B: Agricultural land use suitability analysis using GIS and AHP technique. Computers and Electronics in Agriculture. 2013; 97: 71–82.

[19] Dickson BG, Prather JW, Xu Y, Hampton HM, Aumack EN, Sisk TD: Mapping the probability of large fire occurrence in northern Arizona, USA. Landscape Ecology. 2006; 21: 747–761.

[20] Zheng X, Lv L: A WOE method for urban growth boundary delination and its applications to land use planning. International Journal of Geographical Information Science. 2016; 30(4): 691–707.

[21] Stewart TJ, Janssen R: A multi objective GIS-based land use planning algorithm. Computers, Environment and Urban Systems. 2014; 46: 25–34.

[22] Malczewski J: GIS-based multicriteria decision analysis: A survey of the literature. International Journal of Geographical Information Science. 2006; 20: 703–726.

[23] TSS. Turkey Statistical Service. Population records of the Van Province. 2016.

[24] Özyavuz M, Şatır O, Bilgili BC: A change vector analysis technique to monitor land use/land cover in Yıldız Mountains, Turkey. Fresenius Environmental Bulletin. 2011; 20(5): 1190–1199.

[25] Berberoglu S, Lloyd CD, Atkinson PM, Curran PJ: The integration of spectral & texture information using neural networks for land cover mapping in the Mediterranean. Computers and Geosciences. 2000; 26: 385–396.

[26] Foody GM: Status of land cover classification accuracy assessment. Remote Sensing of Environment. 2002; 80: 185–201.

[27] Lillesand TM, Kiefer RW. Remote Sensing and Image Interpretation. New York: Oxford University Press; 1994.

[28] Voogd H. Multicriteria Evaluation for Urban and Regional Planning. London Pion, Ltd. 1983.

[29] Zadeh LA. Fuzzy sets. Information and Control. 1965; 8: 338–353.

[30] Mosadeghi R, Warnken J, Tomlinson R, Mirfenderesk H: Comparison of fuzzy-AHP and AHP in spatial multi-criteria decision making model for urban land-use planning. Computers, Environment and Urban Systems. 2015; 49: 54–65.

[31] Wang F: The use of artificial neural networks in a geographical information system for agricultural land-suitability assessment. Environment and Planning A. 1994; 26(2): 265–284.

[32] Kalogirou S: Expert systems and GIS: an application of land suitability evaluation. Computers, Environment and Urban Systems. 2002; 26: 89–112.

[33] Cengiz T, Akbulak C: Application of analytical hierarchy process and geographic information systems in land-use suitability evaluation: a case study of Dumrek village. International Journal of Sustainable Development and World Ecology 2009; 16(4): 286–294.

[34] Chandio IA, Matori AN, Lawal DU, Sabri S: GIS-based land suitability analysis using AHP for public parks planning in Larkana City. Modern Applied Science 2011; 5(4): 177–189.

[35] Bunruamkaew K, Murayama Y: Site suitability evaluation for ecotourism using GIS&AHP: A case study of Surat Thani province, Thailand. Procedia Social and Behavioral Sciences. 2011; 21: 269–278.

[36] Javadian M, Shamskooshki H, Momeni M: Application of sustainable urban development in environmental suitability analysis of educational land use by using AHP and GIS in Tehran. Procedia Engineering. 2011; 21: 72–80.

[37] Zolekar RB, Bhagat VS: Multi-criteria land suitability analysis for agriculture in hilly zone: Remote sensing and GIS approach. Computers and Electronics in Agriculture. 2015; 118: 300–321.

[38] Wang Y, Ding Q, Zhuang D: An eco-city evaluation method based on spatial analysis technology: A case study of Jiangsu province China. Ecological Indicators. 2015; 58: 37–46.

[39] Aydi A, Abichou T, Nasr IM, Louati M, Zairi M: Assessment of land suitability for olive mill wastewater disposal site selection by integrating fuzzy logic, AHP and WLC in a GIS. Environmental Monitoring and Assessment. 2016; 188: 59.

4

A Theoretical Framework on Retro-Fitting Process Based on Urban Ecology

Selma Çelikyay

Abstract

Today, natural resources have been destroyed to make developed built environment. Furthermore, this destruction is on behalf of the creation of quality urban environments. It is necessary that dynamics of the planning, design, application processes which build urban environments should be taken into account and firstly these processes should be retrofitted. In this context, approaches, principles, action plans, and application tools from planning to architecture, from architecture to building material, from material to production and consumption technologies based on ecology should be ranked in a hierarchical row and this frame should be adopted. This ecological framework is the theoretical, legislative, administrative base for sustainable settlements, and sustainable and livable cities, in addition to the creation of the quality built environments in which the community has a chance for comfortable and quality living condition. In this chapter, urban ecology and ecosystems of the cities ,which can be embraced as living organisms, are mentioned. With the integrated approach, beginning ecological planning, importance and utilization of eco-technologies also are emphasized in the framework of ecological architecture and ecological urban design based on urban ecology.

Keywords: Ecological planning, ecological urban design, ecological architecture, eco-technologies, urban ecology

1. Introduction

In most of the cities of some countries, due to rapid urbanization, urban growth related to urban population growth has been seen. Rapid growth in built environment and land uses without planning and lack of ecological approach have natural environment make smaller and have risen to ecological and biological destruction on natural resources. When the global population continues to grow, dramatic challenges in the early twenty-first century are those ecosystems are changing from the landscape to the global scales, due to increase in population and urbanization [1]. It is time to consider all decisions from planning strategies to architectural design and usage of technologic tools for retro-fitting ecological condition in the cities. The result to be occurred is to change the view point looking and taking into account or monitoring cities and settlements in which we live.

Adoption of the city as an ecosystem gives rise to protection and consideration of this ecosystem for spatial organizations at every scale and adoption of ecological balance in land use planning. Fictionalize of urban space is a process starting planning. If ecological approach guides not only in planning, but also in urban design and building design process, natural resources can be used without consumption and it can be enabled that people live in a healthy urban environment.

Urban ecology has become an expanding field of research during the last two decades. Various studies carried out on urban climate, soils, flora, fauna, urban habitats, and green space of cities [2]. Urban areas contain many land use types and every land-use type has a distinctive structure from vegetation perspective. The knowledge of characteristic vegetation structure of land-use types can help making a connection between urban areas and surrounding natural areas and taking advantage for studies building up urban open space systems [3].

Recently, although some issues like rational utilization of natural resources and energy efficiency in buildings have been taken place in legal legislation in some developing countries, still there are some issues to be added to the legislative frame and application process in order to decrease ecological footprint on the settlements and built environments, not only consume at minimum level of natural resource but take maximum advantage from them. Transformation of nonrenewable natural resources to the unique capital of socio-economic development and growth process is depend on adoption of the eco-economic policies and utilization of the eco-technologies enabling the life circle and increasing retrieval possibilities of wastes which occur in every steps of consumption process.

There is an increasing need to enable the transformation of our preferences from using nonrenewable resources to renewable energy, non-consumable resources in order to supply fundamental human requirements [4]. This will demand a large data source of ecological planning and design tools, indicators, case studies, and applicators.

For sustainable urbanization, as many scientists have pointed out, growing urban environment depend on human's increasing demands, it is time to consider planning, design, and producing strategies and to measure with indicators ecological footprints, CO_2 emissions, air and water pollution, degradation, deforestation, and unpredictable urban sprawl. To realize monitoring and controlling the issues mentioned above, there is a need for frameworks theoretically constituted and applicable.

2. Ecological approach in organization process of urban space

On the one hand, urban growth and urbanization have caused built environment to expend, on the other hand, built environment growing together with the economic development has negative impacts on the natural resources. From planning perspective, in most of the settlements or cities, natural resources have not been taken into account during urban growth and economic development due to the lack of ecological approach in planning or decision-making process [5].

From planning to design, there are three dimensions of ecological approach for making cities livable. None of them has not priority but all of them should be considered and guide to spatial organization process of urban space for livable and sustainable cities and also sustainable urbanization. Human, nature and built environment trio and the relation among them should guide to planners, urban designers, architects, urban policies in planning process from the beginning to the end based on ecology. From people-oriented approach, basically and

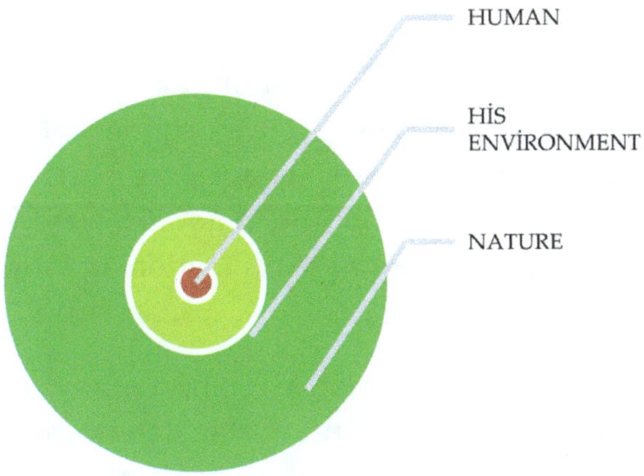

Figure 1. Human in the nature.

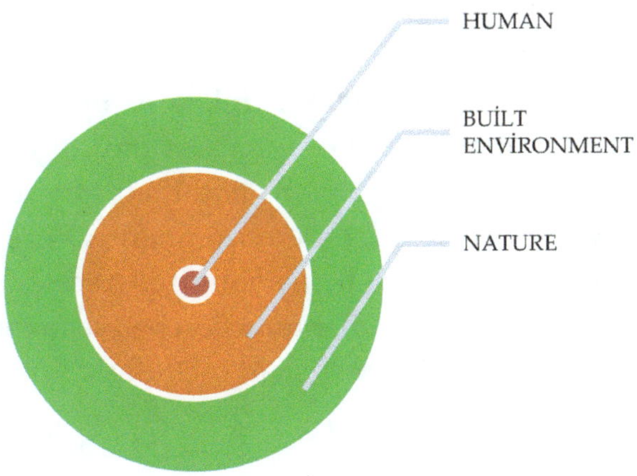

Figure 2. Human and growing built environment in the nature.

schematic presentation on the relation of human and nature is that human is a point in the nature (**Figure 1**).

There is an interaction between human and the nature, both human has an impact on the nature for his vital activities and the nature influences on the human. Once, it was an organic interaction among human, surroundings and nature (**Figure 1**). But negative and devastating effect on the other is coming from human (**Figure 2**). Human changes his environment for housing, recreational, commercial, or industrial activities. This is the beginning of his making nature destroyed when built environment occurs.

The simplified relation between human and built environment surrounded him, shown in **Figure 2**, is everywhere in which human groups or community live. From holistic approach to nature, built environment, which is the footprint of human in the nature, looks as if it is a kind of viral infection covered the whole body. Beginning ecological framework, it is of great importance that understanding human and urban environment from aspects based on ecology is necessary.

2.1. Human ecology

Originally coming from the Greek root oikos meaning "home", the term of ecology was the basis for the field of home economics. Then, it was changed to human ecology by many contemporary researches. Looking at humans as not only social beings, but also biological organism, human ecology theory focuses on interaction between human and their environments; thus, emphasized issues in this theory are utilization and management of resources for human development and sustainability [6].

Ecology is commonly defined as the study of the relation of organisms to their environment [7]. The position of human ecology in this context was determined primarily by the point of view cultivated initially sociologists, particularly to account for certain aspects of American cities in the first quarter of nineteenth century [7]. While the cities of that period were experiencing rapid and troubled growth, and therefore, the focus of human ecology changed direction to a concern with the humans' environment. This aspect added useful content to the issue of spatial analysis [7].

The city is a conspicuous example of a system of relationships among differentiated activities by means of which a population is able to occupy a unit of territory. Although ecology includes the relation between all living organisms and their environment, human ecology is on the relation between only humans and their environment. Quite clearly, therefore, human ecology is a sociological concern [7]. From this point, it is necessary to consider interaction between human and environment and as well interaction between the social system of humans and the rest of the ecosystem, because an ecosystem includes all built elements by humans as parts of the ecosystem [8].

Based on short definitions and approaches about human ecology mentioned above, this interaction and relationship should be taken into account as a system, in decision and policy making process on urban land, so that next generations can live sustainable settlements.

2.2. Urban ecosystem

As outlined by most theorists, an ecosystem consists of three indicative components: humans, their environment, and the interactions between them. In urban ecosystems, high-density population, or the built structure of environment occupies a large place on the land [9]. There is now convincing proof on changing nearly all of the earth's ecosystems by humans [10]. With urbanization becoming a global phenomenon, urban population increased all over the world causing huge pressure on the environment [11]. Land consumption, habitat fragmentation, and biodiversity loss may have negative impacts on society and economic systems on a local scale, impacting urban sustainability [11].

In order to comprehend the impact of urbanization, a clear consideration of the temporal dynamics, although constantly missing, is vital for determination impacts on biodiversity in rapidly urbanizing landscapes [12]. While more than half of the natural resources are used by humans, whereas approximately half of the land cover has been changed by human actions causing important losses of biodiversity [10]. As a result of these activities, most ecosystems can argumentatively be taken into account human-dominated ecosystems; nevertheless, humans are also building different ecosystems particularly for residence as urban ecosystems [10].

Although city is a settlement in which people live and whose population exceeds the certain amount, urban ecosystem is a living organism with specific dynamics and components. If relation and interaction of the elements and components in this organism occur according to ecological principles, it is possible that livable and healthy urban spaces can be built up. Urban lands contain heterogeneous materials with a variety of lands and complex interactions. Excluding water surfaces, the combinations of impermeable surface materials, greenery plants, and soil are essential elements of the urban ecosystem [13]. An evaluated model contains above approaches *"in reference* [13]" which proposes a large content of data for urban ecosystem researches on physical and ecological characteristics of the nature and human disturbances.

Comprehension the relation and interaction between humans and natural components of urban ecosystems are major characteristics of the integrated model combining social qualifications of humans and their establishments [14]. As the emergent phenomena of local scale [15], cities are now recognized as ecosystems, and many of the concepts and theories considered central to ecosystem ecology can be modified for application to urban systems [16]. Long-term environmental changes in urban ecosystems have raised commitments among scientists, researchers across several disciplines, having a determining role in policies, plans, designs, and management strategies to reply to these changes. Integrating ecological researches with urban policies, planning, design, and management strategies is complex, yet it is one of the significant themes and research preferences in landscape ecology [17]. It is necessary that ecologists, social scientists, urban planners, urban designers, landscape architects, and whoever concerns urban issues to use such interdisciplinary knowledge for development of cities and sustainable settlements, they need to work in collaboration with interdisciplinary approaches in order to address this challenge.

2.3. Urban ecology

Urban areas and ecology are contrary to each other theoretically and as to substance, urban ecology connects with them. Urban ecology was originally developed as a part of human ecology in the 1920s by a small but influential group of sociologists at University of Chicago and they identified urban ecology as "the study of the relation between people and their urban environment" [18]. Urban ecology is a scientific field on the issue of effects on organisms, the interactions among them, and the transformation and flux of energy and matter in urban and urbanizing systems [16]. Urban ecology is ecological research conducted in cities, towns, and urban areas [19]. To understand the structure and dynamics of urban systems, they must be recognized as social-ecological systems that integrate socioeconomic drivers and responses with ecological structures and functions [16].

Basically, "urban" is a term identifying the land use on a place. A common comprehension of "urban" a densely populated urban space characterized by multifunctional land uses. This comprehension seems to be more useful for researches on urban ecology; yet, it is difficult to determine ecological thresholds surrounding urban space. Although ecology is a natural science concerned with the distribution and abundance of organisms, the word has many other meanings as well [19]. Urban ecology is not only a view to urban land from the science of ecology but may include concerns from the social sciences. Urban ecology is a research subject forming a continuum from "pure" ecology in the urban space to an integration of ecology and social sciences to investigate urban systems [19]. Today, the sociological approach to urban ecology continues to exist and evolve [18].

An explicit focus of researches about urban ecology is on sustainability, which includes biodiversity and ecosystem services, energy consumption and carbon footprint for climate change studies [20]. Implementation of nature conservation in the cities will be possible by means of arrangements make harmonic natural–cultural and social environment with each other. Although the conflicting approach on ecology of urban areas focused on ecological footprints, and summaries of citywide species richness, contemporaneous ecosystem approaches have begun to integrate physical environment of the nature, including urban climate, hydrology, soils, and energetic issues, and to indicate the necessity for comprehension the social dimensions of urban ecology [9]. Urban ecology has increasingly played a significant leading role to overcome the conflicts faced by urban ecosystems. The major focus in many studies about application of urban ecology was on urban green network and its connection to citizens [20]. Three main goals based on ecology "in reference [9]", "may be achieved in urban areas; Firstly, plant ecology can support to improved comprehension of the structure and function of urban ecosystems. Second, ecological function of urban areas must be increased. Third goal is to take advantages to humans of the vegetation elements of urban lands" [9].

As emphasized by many scientists and researchers, with an in-depth comprehension of ecology in cities, focusing on the ecology of cities is increasingly being more significant. A framework to put ecological information into practice in urban areas requires interdisciplinary approaches for enhancing social welfare and providing sustainability in urban areas.

3. Tools for ecological retro-fitting of urban environment

Ecological approach should accompany to planners, urban designers, architects, landscape architects, builders, from beginning of the planning process to production of the urban elements and building materials [5].

From ecological point of view, to retro-fit our settlements and urban environments in which we live, to reduce ecological footprints on the settlements, it is essential that ecological approach and ecological decision making systematic should be adopted. This systematic begins with ecological planning and contains ecological urban design for urban landscape and ecological architecture and eco-technologies for livable and sustainable cities.

Figure 3 shows the hierarchic succession of the phases from planning to architecture and to the production technologies.

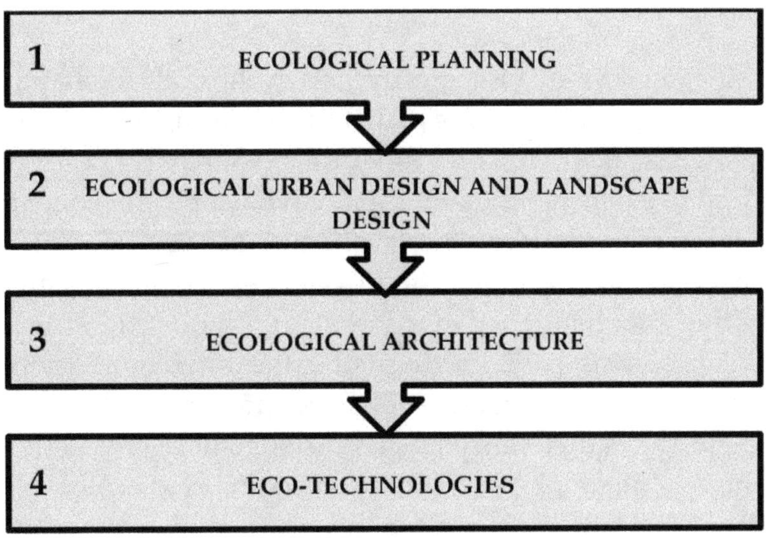

Figure 3. Tools for ecological retro-fitting of urban environment.

3.1. Ecological planning

Planning is a decision-making and spatial organization process connecting with human, nature, and life (**Figure 4**). In addition to this definition, planning also creates a system including connection and interaction among social, natural, and built environment. As the key issue of sustainable development, the principle of "using of the natural resources without exhausting them" was adopted nearly by the whole world and the space planning strategies gained a new dimension which considers natural resources [5, 21–23]. In the world, spatial planning strategies targeting sustainable development of cities gained ecological dimension guiding to both the regional and city planning process [21, 23]. Ecology should be the major theme to be considered from the beginning to the end of planning process [5]. Consideration of natural resources and ecological characteristics of an area is of great importance. In ecological land-use planning process, strategic environmental assessments, impact assessments, and ecological risk analysis are of great importance as basic phases [5, 21, 22].

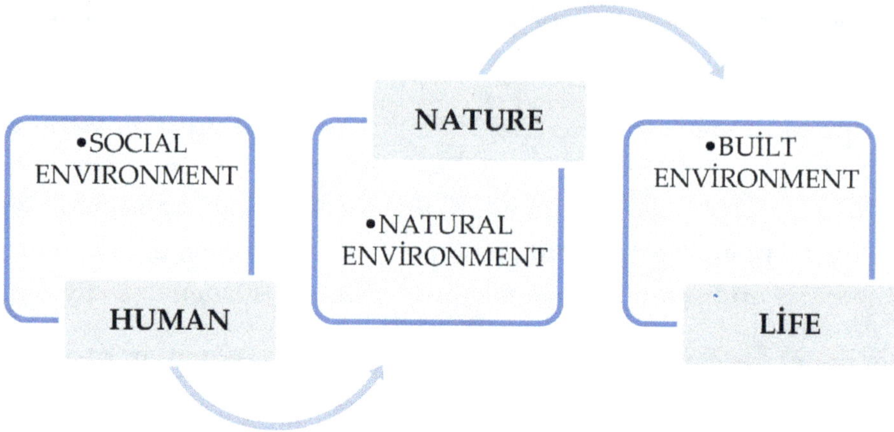

Figure 4. Function of planning.

Land use involves the process of biologically and technically reshaping, converting and managing land for socio-economic benefits [24]. Built environment growing together with the economic development has taken a toll on the natural resources [22, 23]. Overexploitation and utility of resources due to extensive economic growth have now become the common causes of environmental degradation [24]. One the one hand, land-use decisions on the settlements impact economic actions, this also affect the physical environment, land surface, soil, and natural biotopes, on the other hand, negative impacts of these actions on ecological structure cause environmental problems [23]. Furthermore, the interconnectivity of ecosystems is increasingly jeopardized by the uncontrolled development, land-use conversion, tourism activities, and endless disturbances caused by anthropogenic agents [24]. There is a need also for cost-benefit analysis in planning process, so that it can bring ecological benefit, besides of economic development in growth process of the cities. While some of land-use decisions economically benefit in the short run, whereas they cause ecological costs in the long run [21, 23].

Therefore, the land-use decisions should be defined by environmentally sensitive/ecological planning process, during which natural characteristics and potentials of the lands must be examined by ecological threshold analysis [21–23]. The aim of ecological planning was to protect of natural resources and to create ecological cities and settlements. Ecological planning is not only physical planning understanding, but also a planning process which aims sustainability of an integrative urban-nature development [23].

On the one hand, one of the most important approaches in the creation process of ecological cities is to utilize of natural resources efficiently and without consumption of them. On the other hand, one of the basic principles in ecological planning process is to create an integrated green network in the cities. The management of urban lands, namely adding green components to open spaces is an increasingly important subject. Urban biodiversity is essential for residents as recreational areas, and their presence in the neighborhood unit is an appreciated characteristic reflected in property prices [25]. Open and green spaces, which were planned, applied, directed with the ecological, and recreational aims, play an important role for controlling urban growth and shaping urban form, in addition to being important for residents [26]. Open and green lands in the cities can perform their functions if the practices are suited development projects and ecological planning rules.

Geographical information systems (GIS) are major tools for planning in order to make available to spatial outputs of land-use decisions. Thus, increasingly approaches using GIS for visualizing the spatial dimension and extent of change are being employed to ensure these spatial aspects [27].

3.2. Ecological urban design and landscape design

In rethinking the city as a "human ecosystem", urban design is emerging as one tool for adaptation as we face complex, wide-ranging issues from food security to sea-level rise. The practice of urban design is interdisciplinary in nature, often claimed by several professions and residing somewhere at the intersection of urban planning, architecture, and landscape architecture.

Urban design process is a period in which planning decisions change to applications. In this context, urban spaces are created adding third-dimensional to two-dimensional planning decisions. Urban design is both an interface and a bridge between planning and architecture. The aim in urban design process including designing of the streets, avenues, squares, public spaces, green areas, plants, and surface waters was to create healthy living areas, besides fictionalize of an aesthetic environment. The main issue of ecological urban design process is urban ecology. Green areas, plants, and water surface are ecological components of urban landscape surrounding buildings. With the ecological urban design and ecological landscape design approaches, urban design, landscape ecology, the knowledge of policy makers, and the demands of neighborhood dwellers are integrated for creation of new patterns of greenery areas, buildings, and land covers.

Recently, due to rapid urbanization, green infrastructure within the built environment has gained importance. Thus, most of the cities in the world have been designed adding green areas to public and open spaces. Plants as green elements make environment relaxing, green, and aesthetic in addition to healthy for city community. Linking plant ecology with urban design and landscape design can help to take advantage of urban design projects as useful tools for ecological researches [9]. Plants at the parcel scale including open spaces, multifunctional land uses, and pedestrian walkways are of great importance, namely in high-density neighborhood designs to enhance physical performance and improve citizens' life [28].

When the world's cities are becoming increasingly crowded and polluted, green spaces in urban areas provides a lot of ecosystem services that could help struggle with many urban ills and improve living for citizens, namely their health [29]. From urban ecology aspects, green and ecological networks have been important elements of urban planning in recent years and have been organized based on ecological and social services to provide regulated climate, recreational occasions, environmental conservation and biodiversity, while they also supply social and psychological necessities of the community [30]. While residential landscapes have been taken into account for their aesthetic value, recently, residential landscapes, especially trees, have been interested, due to their role in ecosystem services, such as reducing air pollution, reducing energy use, and regulating microclimatic conditions in the cities [31].

There is a need to gain experience to make a bridge between human and nature from urban and landscape design aspects. Our neighborhoods and cities, as designed mess, face to the lack of practice of design, which concerns comprehension of ecology [32]. As emphasized by Ryn and Cowan "*in* [32]", "until our everyday activities keep ecological integration by design, their cumulative effect will keep to be destroying". Ecological planning, ecological urban design, and landscape design provides a framework for retrofitting and redesigning urban environments, streets, avenues, landscapes, buildings, and cities with the natural components, besides of physical elements. Urban designers imagine the future of the city, and point out their architectural and infrastructural perspectives from the neighborhood to the regional scale [9]. These approaches are essential principles for profession experts on architecture, landscape architecture, urban planning to plan and design in urban areas.

3.3. Ecological architecture/ecological design

Beginning planning process and keeping on urban and landscape design at urban scale, ecological principles, and indicators should be also taken into account at building design. Ecological principles in spatial planning can be realized when they are kept on architecture. This approach based on ecology is essential either for sustainability or sustainable urbanization.

Ecological design was defined "as any form of design that minimizes environmentally destroying effects by integrating itself with living processes" by Ryn and Cowan "*in reference* [32]". This integrative definition addresses ecologically responsible design, which is not only a style, but also a form integrated with nature. In the context of the definition, "ecological architecture is eco-design approach to policies that an enhanced thought systematic", enhancing the conservation and improvement of microclimate, the new buildings for green areas by improving the energy and ecological principles of old buildings, and is located in the renewal of ecological architecture thought systematics [33]. Based upon the principle of ecological architectural design, it discusses architectural ecological design strategy from water saving, energy saving, earth resource saving, humanistic environment, and greening ecological environment, so as to enhance the sustainability of urban development and realize the sustainable development of architectural industry.

With the ecological/sustainable design approach, the effects on the environment of the construction materials used in building should be examined [34]. The scope of ecological design embraces the study of architects rethinking their preferences of building materials [32]. The production process of building materials has negative effect on the nature; "harvesting trees could result in deforestation; mining mineral resources destroys the nature and causes environmental pollution" [34].

As it should be in both planning and urban design-landscape design process, one of the ecological indicators of architecture is to integrate of architecture and greenery as a part of urban green network. Some researches on sustainability of cities have promoted the application and protection of green components in the urban context. At this point, green architecture and green planning are of great importance. The benefits of plants, which will improve the climate and ventilation, reduce energy and water usage, are not just environmental but recreational, aesthetic, and emotional. In the last 10 years, a lot of studies about the full benefits of plants in building and the role they play in the ecology of cities were conducted. In addition, many indicators were evaluated by researchers for measuring of ecological benefits of green components in architecture.

With the multiple expressions based on "design with nature" approach, recently, "ecological accounting" has become a major force in architecture and construction. The guidance of ecology in design, it creates new aspects to design to minimize energy and materials use, reduce pollution, improve climate, preserve habitat, and to increase community well-being.

3.4. Eco-technologies

Although technologic improvement and economic development have increased in most countries, the numbers of livable cities and healthy environments have decreased. It is time to exchange damaging behaviors of the human and to reduce ecological footprints on the earth and on the settlements [5]. The cities experienced unpredicted economic development in recent years, have faced severe challenges caused by ecological and environmental degradation due to over-consumption of natural resources. Although ecology and technology look as if they are contrary to each other, taking into account the connection between them makes cities livable and sustainable.

Despite susbstantial socioeconomic achievements, concerns are growing over water availability and pollution, land degredation, and depletion of exhaustible resources [35]. Rapid urbanization associated with economic development is thus considered to be unsustainable, and supporting the ambitious goals of building an economically livable and environmentally sustainable society is difficult over the long term [35]. Urbanization is an integral component of economic development and civil evolution, as observed in both developed and developing countries. Energy consumption and CO_2 emissions are a direct result of the urbanization and industrialization process. Urbanization planning must include purposeful CO_2 emission control through comprehensive and scientific design [36].

Ecological engineering and eco-technology can be defined "as the design of human society with its natural environment for the benefit of both" [37]. Eco-technology can also be identified "as the usage of technological methods for environmental management so as to minimize damage to the environment" [38]. However, as technology uses methods causes harmful impact on environment, eco-technology uses remediation techniques with the aim of a greenery environment and environmentally friendly products or processes [38].

The protection of natural environment resources, which are impossible regain, is bound up with eco-economic policies, adoption of eco-technologies and usage of them. Environmental consequences of productions and services have also become a significant theme for enterprise management. Resent approach on enterprises' environmental precautions aims environmental performance of production process, especially, through eco-innovation [39]. In order to minimize environmental effects of a production and to realize its life cycle, design, and post-design processes should be organized in accordance with ecological principles.

4. Conclusion

In order to retrofit settlements and cities in which we live and make them more livable and sustainable, it is necessary that our point of view and approaches related to urban spatial arrangements should be retrofitted and be gained ecological dimension. Ecology should be the key issue in the spatial planning and design process at every scale.

This chapter proposed a theoretical framework on retrofitting process based on urban ecology for making cities, settlements, and the whole environment livable and sustainable. From planning to design and to production technologies of the building and landscaping materials, adoption of throughout process based on ecology and ecological principles is essential so that habitats for the next generation to live in the earth have not been destructed.

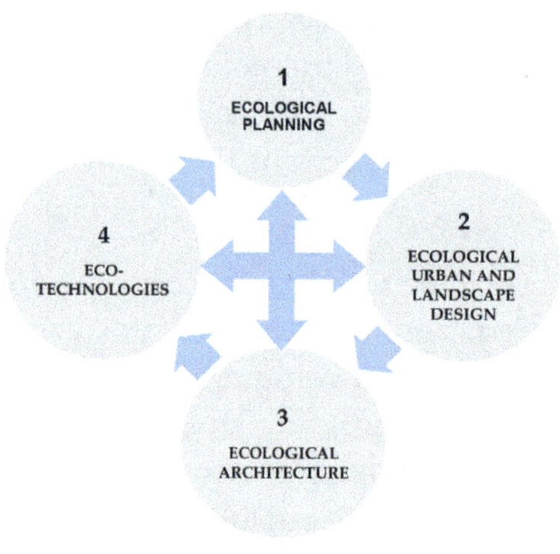

Figure 5. Ecology-based retro-fitting cycle in the framework.

With the guiding of the hierarchic succession on both creating and retro-fitting process urban environment shown in **Figure 3**, decisions, strategies, actions and studies conducted in every phase should consider the other phases' and should be integrated to each other. Thus, the framework emphasized in this chapter is an ecology-based cycle which contains integrated tools for ecological retro-fitting of urban environment (**Figure 5**).

Following this theoretical framework, it may not be remove ecological footprint of the human on the earth, but it would be reduced considering ecology from planning for human and making decisions for land uses in the urban areas to production processes of building and industrial materials. Thus, ecological and biological destruction on natural resources due to rapid urbanization would be stopped. With the sustainability of natural resources and ecologically balanced conservation-utilization strategies, livable and sustainable urban environment and cities can be realized.

As a summary of the retro-fitting framework based on ecology, the following checklist of which some items suggested by author [5], can be used for a guide to sustainability of natural and ecological resources for sustainable cities:

- At the beginning, ecological holistic approach and "planning–design–production based on ecology" should be adopted for integrated action plan
- As an inventory, data regarding natural resources in the planning area should be stored
- Ecological planning is the first process to determine suitable lands for new land uses
- As a strategic planning, ecological planning should begin with strategic environmental assessments
- Sensitivity of natural resources to negative impacts and possible effects caused by current land uses and activities on the lands should be investigated by ecological risk analysis
- With the help of ecological threshold analysis, new land uses suitable for natural ecological resources should be determined
- The balance between protection and utilization of the natural resources should be considered
- With the aim of sustainability of an integrative urban-nature development, an integrated green network in connection to the citizens should be created in the cities
- In urban design and landscape design process based on ecology, the demands of neighborhood residents should be combined to create a new pattern of vegetation, buildings, and surface covers
- Ecological architecture/ecological design should be major force on building
- From the planning to the building and production of the building materials, ecology based approach should guide to planner, architects, landscape architects, builders, and applicators
- Eco-technologies should be used for production of from building to industrial materials
- Environmentally sensitive local administration and environmentally sensitive community are also essential to create sustainable and livable environments, eco-cities, eco-tech settlements, or to retro-fit urban environments

From spatial planning and design point of view, eco-cities, eco-tech cities, smart cities, which are final phase of hierarchic succession of integrated retro-fitting tools, are ecology based settlements in which natural resources are not exhausted, technology is used for ecology (both human ecology and urban ecology), and there is a sustainable balance between economy and ecology.

Eco-tech has great influence on the construction of environment. The utilization of eco-tech is the necessary means of architecture to the aim of sustainable development [40]. Ecological and technological design of new comfortable, healthy, environment-friendly, minimum carbon consuming, self-efficient settlements contribute to urban sustainability [41]. From design point of view, eco-technology can be identified as the design of human society with its natural environment for the benefit of both. Eco-tech planning/design for eco-cities considers energy, environment and ecology for human wellbeing.

The principles of eco-tech city can be explained as the following: "Eco-tech city, when planned in collaboration with local investor, local management and technical team, can decrease ecological footprint of that city by using high performance ecological building and city technologies" [42].

Among the various types of "eco-city" initiatives currently underway across different parts of the world, this tension is arguably particularly pronounced in the case of entirely new, large-scale cities. According to a recent global survey, of the 80 or so recent eco-city initiatives-which include the retro-fitting of existing cities and urban expansion through variously sized "eco-districts" and "eco-towns"—there are a dozen or so brand new cities in the process of being realized [43].

Author details

Selma Çelikyay

Address all correspondence to: scelikyay@bartin.edu.tr

Department of Landscape Architecture, Faculty of Forestry, University of Bartın, Bartın, Turkey

References

[1] Steiner F. Frontiers in urban ecological design and planning research. Landscape and Urban Planning. 2014;125:304–311. doi:10.1016.2014.01.023

[2] Rebele F. Urban ecology and special features of urban ecosystems. Global Ecology and Biogeography Letters. 1994;4(6):173–187. doi:10.2307/2997649

[3] Deniz B., Eşbah H., Küçükerbaş E.V., Şirin U. Kentsel alan kullanımlarındaki vejetasyon yapısının analizi. Ekoloji. 2008;17(66):55–64.

[4] Ryn S.V.D., Cowan S. Ecological Design Tenth Anniversary Edition. Island Press; 1996. 239 p.

[5] Çelikyay, S. Ecological planning for livable and sustainable cities, a case study for Bartın city. In: Proceedings of the International Conference on

Sustainable Development of Contemporary City; 6–7 December, 2013; Bakü, Azerbaijan. pp. 40–48

[6] Bubolz M.M., Sontag M.S. Human ecology theory. In: Boss P.G, Doherty W.J., LaRossa R., Schumm W.R., & Steinmetz S.K, editors. Sourcebook of Family Theories and Methods: A Contextual Approach. Springer; 1993. pp. 419–450. doi: 10.007/978-0-387-85764-0_17

[7] Hawley A.H. Human Ecology: A Theoretical Essay. The University of Chicago Press, Ltd.; 1986. 168 p.

[8] Marten G.G. Human Ecology-Basic Concepts for Sustainable Development. Earthscan; 2001. 256 p. ISBN: 9781853837142

[9] Pickett S.T.A., Cadenasso M.L. Linking ecological and built components of urban mosaics: an open cycle of ecological design. Journal of Ecology. 2008;96:8–12. doi: 10.1111/j.1365-2745.2007.01310.x

[10] McIntyre, N.E., Knowles-Yánez, K., Hope, D. Urban ecology as an interdisciplinary field: differences in the use of "urban" between the social and natural science. Urban Ecology: An International Perspective on the Interaction Between Humans and Nature. 2008;49–65. doi:10.1007/978-0-387-73412-5_4

[11] Salvati L., Sateriano A., Rontos K. Towards an indicator of urban centrality? Exploring changes in present and resident population (1911–2011) in Greece. Ecological Indicators. 2016;61:188–192. doi:10.1016/j.ecolind.2015.09.011

[12] Ramalho C.E., Hobbs R.J. Time for a change: dynamic urban ecology. Trends in Ecology and Evolution. 2012;27(3):179–188. doi:10.1016/j.tree.2011.10.008

[13] Ridd M.K. Exploring a V-I-S (Vegetation Impervious System) analysis through remote sensing: comparative anatomy for cities. International Journal of Remote Sensing. 1995;16(12):2165–2185. doi:10.1080/01431169508954549

[14] Pickett S.T.A., Burch W.R., Dalton S.E., Foresman T.W., Grove J.M., Rowntree R.A. Conceptual framework for the study of human ecosystems in urban areas. Urban Ecosystems. 1997;1(4):185–199. doi:10.1023/A:1018531712889

[15] Alberti M., Marzluff J.M., Shulenberger E., Bradley G., Ryan C., Zumbrunnen C. Integrating humans into ecology: opportunities and challenges for studying urban ecosystems. BioScience. 2003;53(12):1169–1179. doi:10.1641/0006-3568(2003)053

[16] Cadenasso M.L., Pickett S.T.A. Urban Ecology. Encyclopedias of the Natural World, Encyclopedia of Theoretical Ecology. University of California Press; 2012(4).pp.765–770 ProQuest Ebrary [Accessed 2016-01-19]

[17] Musacchio L.R., Wu J. Collaborative landscape-scale ecological research: emerging trends in urban and regional ecology. Urban Ecosystems. 2004;7:175–178. doi:10.1023/B:UECO.0000044034.55695.bd

[18] Wu J. Urban ecology and sustainability: the state-of-the-science and future directions. Landscape and Urban Planning. 2014;125:209–221. doi:10.1016/j.landurbplan.2014.01.018-0169.2046

[19] Niemela J. Is there a need for a theory of urban ecology? Urban Ecosystems. 1999;3(1): 57–65. doi:10.1023/A:1009595932440

[20] Breuste J., Qureshi S., Li J. Applied urban ecology for sustainable urban environment. Urban Ecosystems. 2013;16:675–680. doi:10.1007/s11252-013-0337-9

[21] Çelikyay S. Determination of land uses by ecological threshold analysis, a study on Bartın case [thesis], Yıldız Technical University; 2005.

[22] Çelikyay S. Determination of land use by ecological threshold analysis. In: Proceedings of the International Congress on Information Technology in Agriculture (ITAFE'05); Food and Environment. October 2005; Adana. pp. 419–423

[23] Çelikyay S. Research on new residential areas using GIS. In: van Leeuwen J.P., Timmermans H.J.P., editors. Innovations in Design & Decision Support Systems in Architecture and Urban Planning. Springer; 2006. pp. 221–233.

[24] Leman N., Ramli M.F., Khirodin R.P.K. GIS-based integrated evaluation of environmentally sensitive areas (ESAs) for land use planning in Langkawi, Malaysia. Ecological Indicators. 2016;61:293–308. doi:10.1016/2015.09.029

[25] Niemela J. Ecology and urban planning. Biodiversity & Conservation. 1999;8:119–131. doi:10.1023/A:1008817325994

[26] Güleç R. Kastamonu yeşil kuşak planlaması [thesis]. Gazi University; 2003.

[27] Young H.C., Jarvis P.A. A simple method for predicting the consequences of land management in urban habitats. Environmental Management. 2001;28:375–387. doi: 10.1007/s002670010230

[28] Jackson L.E. The relationship of urban design to human health and condition. Landscape and Urban Planning. 2003;64(4):191–200. doi:S0169-2046(2)00230-X

[29] Wolch J.R., Byrne J., Newell J.P. Urban green space, public health, and environmental justice: the challenge of making cities 'just green enough'. Landscape and Urban Planning. 2014;125:234–244. doi:10.1016/j.landurbplan.2014.01.017

[30] Ignatieva M., Stewart G.H., Meurk C. Planning and design of ecological networks in urban areas. Landscape and Ecological Engineering. 2011;7:17–25. . doi:10.1007/s11355-010-0143-y

[31] Fissore C., Hobbie S.E., King J.Y., McFadden J.P., Nelson K.C., Baker L.A. The residential landscape: fluxes of elements and the role of household decisions. Urban Ecosystems. 2012;15:1–18. First online: 6 July 2011. doi:10.1007/s11252-011-0189-0

[32] Ryn S.V.D., Cowan S. An introduction to ecological design (1996). In: Ndubisi F.O., editor. The Ecological Design and Planning Reader. Island Press; 2014. pp. 191–202. doi:10.5822/978-1-61091-491-8-19

[33] Tönük S. Bina Tasarımında Ekoloji. Yıldız Teknik Üniversitesi Yayını; 2001. 133 p. YTÜ.MF.DK-01.0628.

[34] Çelebi G. Environmental discourse and conceptual framework for sustainable architecture. Gazi University Journal of Science. 2003;16(1):205–216.

[35] Gao J., Tian M. Analysis of over-consumption of natural resources and the ecological trade deficit in China based on ecological footprints. Ecological Indicators. 2016;61:899–904. doi:10.1016/j.ecolind.2015.10.044

[36] Zi C., Jie W., Hong-Bo C. CO_2 emissions and urbanization correlation in China based on threshold analysis. Ecological Indicators. 2016;61:193–201. doi:10.1016/J.ecolind.2015.09.013

[37] Mitsch W.J., Joergensen S.E. Ecological Engineering: An Introduction to Ecotechnology. John Wiley and Sons; 1989. 472 p.

[38] Gianetti B.F., Bonilla S.H., Almedia C.M.V.B. Developing eco-technologies: a possibility to minimize environmental impact in Southern Brazil. Journal of Cleaner Production. 2004;12(4):361–368. doi:10.1016/S0959-6526(3)00033-7

[39] Kobayashi H., Kato M., Maezawa Y., Sano K. An R&D management framework for eco-technology. Sustainability. Special Issue "Innovation and Environmental Sustainability". 2011;3:1282–1301. doi:10.3390/su3081282

[40] Yang Y. Eco-tech and architecture design. Journal of Chongqing Jianzhu University (Social Science Edition). 2000;3:017.

[41] Ercoşkun Ö.Y., Karaaslan Ş. Ecological and technological cities of the future. Megaron Yildiz Technical University Faculty of Architecture .E-Journal. 2009;3(3):283–296.

[42] Ercoşkun Ö.Y., Karaaslan Ş. Guidelines for ecological and technological built environment: a case study on Güdül-Ankara, Turkey. Gazi University Journal of Science. 2011;24(3):617–636.

[43] Joss S., Molella A.P. The eco-city as an urban technology: perspectives on Caofeidian International Eco-City (China). Journal of Urban Technology. 2013;20(1):115–137. doi: 10.1080/10630732.2012.735411

5

The Analysis of Turkish Urban Planning Process Regarding Sustainable Urban Development

Okan Murat Dede

Abstract

Sustainable urban planning is the recent phenomenon arising with the sustainability concept. Urban areas are the main bodies where sustainable policies can be applied and sustainability criteria have to be tested in. Urban life is equal to the social system in the contemporary world. Urban sustainable development becomes crucial and this condition can be obtained with urban planning. In this respect, sustainable urban planning becomes a crucial factor to maintain sustainability. Main problem is how to adapt existing planning procedures and processes into sustainable-based urban planning. In this chapter, it is analyzed how Turkish existing urban planning process can be adapted to sustainable urban planning without a radical paradigm shift in the Turkish planning system.

Keywords: participation, rational comprehensive planning model, sustainability indicators, sustainable urban planning, urban planning process in Turkey, governance, urban plans

1. Introduction

The main aim of this chapter is to define how Turkish planning process, as a rational comprehensive top-down one, adopts to the necessities of sustainable urban planning. It is obvious that most of the academic research focus on the necessity of changing the structure of existing planning systems as a tool for maintaining urban sustainability regarding the land use, built-up environment, and infrastructure. It is the subject of debate that conventional planning approaches become insufficient to ensure urban sustainability. Strategic planning concept, participatory planning, and community-driven planning are all examples and parts of new planning approaches adopted to urban planning process to ensure sustainability of urban settlements.

Planning, especially the "urban planning" concept, has a potential of playing a vital role for ensuring sustainability aim for the future of cities, in other words "sustainable urban development". Especially, the planning structure and planning processes are important variables to maintain sustainability of settlements. Urban planning is the most important tool or factor to maintain sustainable urban development. Characteristics and various dimensions of sustainability concept have to be coherent and compatible with the urban planning dynamics and urban planning processes and techniques.

In order to obtain the sustainable urban development, several models on sustainable urban planning have been offered since 1980s. Planning structure and sustainable urban development together have become one of the most important academic and research subjects since the end of the twentieth century. Especially, new planning initiatives address the necessity of changing traditional comprehensive master planning. In 2001, Naess [1] discussed that as rational comprehensive planning is goal oriented, it is appropriate for sustainable planning, but there are also some shortcomings. There are debates that advocacy and collaborative types of planning can best fit to overcome those shortcomings [2].

Rational comprehensive planning approach is the one which is used most frequently in the world. However, it is criticized that rational comprehensive planning method cannot answer the necessities of sustainable urban development. It is stated that existing urban planning system based on aiming at development goals and prepared by an authority cannot be able to maintain appropriate land use built-up environment and infrastructure in a contemporary world within a changing era [3].

The main aim of this chapter is not to criticize various planning approaches with regard to sustainable urban planning but to criticize how rational comprehensive approach can be adopted to the sustainable planning with reference to the Turkish spatial planning system.

The Turkish planning system is selected to refer the subject because, first of all, the difference between developed and developing countries has to be identified. İf it is a fact that the urbanization concept is simply related to the ratio of population living in urban areas, than it can be easily accepted that developed nations (Japan, North America, and Europe) had almost completed their urbanization as over 80–90% of their population living in urban areas and the increase in the population rates is becoming slower and slower. However, they confront with the mass number of refugees as well as gentrification seems another important urban problem area in front of developed countries. On the other hand, urbanization is still an important factor in developing countries like Turkey. Despite the decreased rates, rural–urban migration is still continuing and urban areas are expanding and developing rapidly against natural environment.

For these reasons, urban planning becomes more important tool to guarantee sustainable urbanization in developing or less developed countries as Turkey is among one of these countries. Turkey is a country, living traditionalism, modernism, and postmodernism at the same time. Turkey's urban planning process is a conventional comprehensive master planning approach, and it is a subject of debate that whether this planning process can be able to ensure sustainable urban development or not because at every passing moment urban built-up environment and natural environment are deteriorating and even ordinary people in the street suffering from these unfavorable conditions.

Especially, Turkish urban plans are mostly executed on design desks, now on computers, and unfortunately, these plans have no chance to be applied because spatial decisions concluded always become the subject of the related Courts as there are a lot of discrepancies about these plans. Most of the planning decisions in these urban plans have a way on courts instead of application.

Depending on this type of frame of reference, after the priorities of the urban planning with regard to sustainability characters are explained, criticism of the Turkish planning system is given and how can this planning system can be adapted to the sustainable urban planning is discussed.

2. Sustainability concept and sustainable urban development

The history of sustainability concept dates back to 1970s. Sustainability concept arose when environmental degradation and economic bottlenecks were increased and developed countries started suffering from these conditions. Therefore, it becomes a necessity being aware of the common future of the humankind. The concept of "Sustainable Development" gained importance after the Work of the World Commission on Environment and Development known as the Brundtland Commission in 1987. The final product is known as the Brundtland Report. This is also the first milestone that sustainability concept starts to reflect itself on urban planning. Sustainable development is a continuing and balanced development model. As cities are the basic elements of contemporary civilization, it can easily be claimed that sustainability concept is an urban-based concept. Sustainable urban areas are the key factors for the success of sustainability approach. This situation is emphasized and gained more importance after the Earth Summit or the UN Conference on Environment and Development in Rio in 1992 and Habitat II Conference held in Istanbul in 1996. After the Habitat II Conference, sustainable development became a necessary strategy that countries have to adopt as a requisite.

"Local Agenda 21 accepted in Rio Conference has a special importance that it could be accepted as the primary mechanism for the application of sustainable development at the local level. This shows that urban areas have become the focus of sustainable development policies starting from the 1990s" [2].

Substantially, the sustainability concept can be adapted to all components, sectors, and institutions of any society. Sustainable city, sustainable economy, sustainable mining, sustainable education, and sustainable environment are some of the examples of this statement. However, sustainable development has a framework that contains five stages as political and supervisory (related to decision making, participatory processes, use of resources, etc.), physical (spatial relations, land use, etc.), environmental (eco systems, artificial urban systems, land, air, water, etc.), economic (production, consumption, employment, etc.), and social (equity, security, life quality, etc.) stages [4].

Sustainable development and sustainable society are urban-based concepts. Though there is a concept of rural sustainability in the literature, it is not wrong to assimilate the sustainable development concept within urban sustainable development as the basic determinants of the civilization are the urban concept and urban life. In addition, urban areas are against the nature and natural ecology as built-up environment of cities invades natural land in a very fast manner. Even you built up a cabin in the woods, you can give damage to the microenvironment and ecology on the area you built up your cabin. The concept of city and urban systems is naturally against the ecological sustainability concept. That is why the urban areas have become the key factors of the sustainable development. Not only the environmental and ecological concerns are the focus of the problems, but also economic and social problems arise from urban systems and mechanisms become the source of global anxiety of future. Especially, as cities of developing world are overgrowing, environmental, social, and economic problems are not remaining as local problems but they are seen as global problems and tried to be solved. In this respect, sustainable urban development takes place at the heart of the sustainability concept. Sustainability of urban areas becomes the key factor in sustainability debates.

Definitely, planning is the main director of urban development and urbanization. Planning not only deals with the physical development of the cities but also affects the social, economic, and cultural future. For this reason, a sustainable urban planning model is very essential for the sustainable urban development and a sustainable society.

3. Factors come into prominence with sustainable urban planning

A new debate arises as how the existing planning processes and methodology fulfill the achievement of obtaining sustainable urban development target or is there a need for new planning paradigm shift with regard to increasing sustainability discussions.

Our discussion is based strategic thinking, inclusive decision-making, governance, participation, monitoring, and sustainability indicators. Especially, these factors are interrelated and interwoven.

Strategic thinking widely depends on and takes its roots from the Urban Strategic Planning Process. This is the planning process offered by UN in the series of publications on "Inclusive and Sustainable Urban Planning: A Guide for Municipalities" [5]. Especially, "strategic planning" is a very wide concept and it represents the adoption of management type of private sector business planning to public planning concept such as urban planning. Indeed, it is not the consequence of only sustainability debates but it is mainly related to the privatization, globalization, and deregulation efforts within the socioeconomic systems. This planning approach is an inclusive, strategic, and action oriented [5].

UN offers the following four phases of urban strategic planning [5]:

a. Urban situation analysis consists of stakeholder analysis, urban situation profiling, urban situation appraisal, investment capacity assessment, and consolidated urban diagnosis.

b. Sustainable urban development planning consists of urban consultations, drafting the strategic urban development plan (SUDP), and approval and adoption of SUDP.

c. Sustainable action planning consists of drafting action plans, local resource mobilization, and public–private partnerships.

d. Implementation and management of projects consisting of project design, management and coordination, and monitoring and accounting reporting.

According to these phases, strategic planning is different from rational comprehensive planning process as it is a bottom-up approach and depends on project making.

Also, it is a dynamic process so that the participation concept is crucial. The urban strategic planning process is based on participatory decision-making approach that all stakeholders involve the plan-making process at any stage. Participation is a wide concept and becomes a crucial factor for sustainable character of an urban plan. Especially, participation contains both citizen participation and participation of several institutions in the planning process. It can be formed as a passive participation of citizens as contribution to questionnaires and surveys, semiactive participation held in meetings, or active participation in which people come together in planning workshops.

In UN's publication on ``Inclusive and Sustainable Urban Planning: A Guide for Municipalities," participation is defined as follows [5]:

- Information (one-way communication in which citizens are informed).
- Consultation (two-way communication in which stakeholders have an opportunity to have suggestions and concerns for the suggestions of other people).
- Consensus building (stakeholders interact to understand each other and arrive at negotiated positions).
- Decision making (expression of full commitment).
- Risk sharing (to take risks all together).
- Partnership (to be at equal status and have a common goal).
- Self-management (stakeholders take the responsibility from beginning to the end).

Participation in the planning process is closely related to the concept of governance. Sustainable development requires well-shaped governance. Furthermore, sustainability itself is defined as a fundamental principle of good urban governance [5]. The sustainability concept emphasize on cities as actors to take balance on the environmental, social, and economic needs of present and future needs. To relate this fact with governance, it means that all individuals and public and private institutions come together to plan, organize, and arrange all the common works of an urban area.

Governance is a broader concept than administration or government. If governance is the structure, then it is easily claimed that elitist planning decisions come from the authority beyond will not be the case for sustainable urban planning. A participatory democratic decision-making process is seen as the requirement of the sustainable planning that governance is the key in the process.

Another important subject is the inclusive decision-making strategy. This is defined as a strategy where norms of good governance are put into practice [5]. Inclusiveness is related with participatory decision-making processes, equal opportunities, safeness, information clarity, equal access to urban services, and consideration of urban poor and marginal groups.

Another determinant and essential factor of sustainable planning is the monitoring process after plan making. Though it is not included in the plan-making process, it has a very crucial meaning for the proper implementation of the plans. When sustainability is considered, monitoring stage is used to test whether primarily defined sustainable goals are accomplished or not.

These goals are highly related with sustainability indicators. Another factor for the sustainable urban planning is the integration of sustainability indicators into the planning process. Rosales and Yazar insisted on the importance and necessity of these indicators in planning. Rosales [6] defines them as ex-ante tools in urban planning. These indicators are classified as environmental, economic, and social. These indicators were first discussed in first Aalborg Conference in 1994 [7] and emphasized that they ought to take place within urban planning in the second Aalborg Conference in 2004 [8].

4. A brief history of Turkish spatial planning system

In this part, the focus is on the basic characteristics and application of a rational comprehensive planning method instead of a detailed history of urban planning and legal aspects of spatial planning in Turkey as it has a very changing structure.

History of the Turkish spatial planning system dates back to the second half of the nineteenth century of the Ottoman period. These efforts are known as spatial arrangements rather than an official urban planning. This situation is also true for the 1920s and 1930s; the first years of the new Turkish Republic. Urban spatial planning was first institutionalized after the acceptance of the first Development and Zoning Law in 1956 numbered as 6785. This date was important as urbanization in Turkey was really accelerated after 1950 hand in hand with industrialization efforts.

As masses of people from rural areas started to migrate to big cities, such as Ankara and İstanbul, new urban planning arrangements and efforts on spatial planning were started with the first Development and Zoning Law accepted in 1956. This law gave way to a rational comprehensive planning approach for the Turkish spatial planning system. Also regarding the whole planning concept; "planned era" was started in 1960 at the country level with the establishment of State Planning Organization. (Note: It was transformed into Ministry of

Development in 2012.) The starting point of the planned development is the Main Law accepted in 1961. After this period, the Five Year Development Plans were started to be prepared for national and regional progress. First plan was accepted in 1962. It was a development plan for the period of 1963–1967. These Five Year Development Plans were organized to be implemented by yearly programs. Though there were no strict decisions on urbanization in successive Five Year Development Plans, there were decisions about urbanization issues and urban planning related to the conjuncture of the period they were applied.

Return to the first Development and Zoning Law accepted in 1956, the planning authority was chosen as municipalities. The law numbered 6785 was not able to prevent pseudourbanization as it was only about the physical planning dimension [9]. This law was replaced by the law numbered 1605 in 1972. The main difference was the fact that authority was given to central authority instead of municipalities. With rapid urbanization, this law and planning efforts became insufficient and the law numbered 3194 was accepted in 1985 instead of the existing one. This law is the current operative law with some changes until it has been accepted.

This law clearly defines and suggests a rational comprehensive urban planning aspect. After accepted, it was criticized as how to engage participation and how macrodecisions are taken, and how policies will be defined and applied to urban areas and urban space [9].

Several directives and bylaws were accepted until 1990s in order to organize and adapt spatial planning to the conjuncture changes and fast-changing characteristics of the society and cities. A new paradigm shift for whole social and economic issues becomes the focal point of academic and political debates. Globalization and effects of information technologies are the main determinants of the new paradigm shift. In addition, the deterioration of the environment and steadily decreasing natural resources are the other important milestones in this paradigm shift. Regarding this part, the last regulation about making spatial plans that was accepted in June 2014 called "Regulation on Making Spatial Plans" is insisted on. The coming part is largely based on the main principles and processes of the Turkish spatial planning system depending on the related ``Regulation on Making Spatial Plans."

5. Turkish current spatial planning process

The Turkish urban planning system is a rational comprehensive one with a strict hierarchy of various kinds of plans varying from strategy plans to urban design stage. Though urban design is appreciated at the Project level, provisions concerning urban design are added to the legislation. Especially, the implementation of the urban decisions is shown at the scale of Implementation Plan level.

"Regulation on Making Spatial Plans" accepted in 2014 clearly defines the hierarchical system of the Turkish planning system [10]. This hierarchy shows the comprehensive character of the planning system as the hierarchy of the plans is one of the most important characteristic of the rational comprehensive urban planning.

Regarding general planning hierarchy, various plan types can be listed as follows:

– Country Plans

– Regional Plans

– Spatial Strategy Plans

– Environmental Order Plans

– Master Development Plans

– Implementation Plans

Especially, Country Plans and Regional Plans are socioeconomic in character while the spatial strategy plan is in between socioeconomic character and high physical plan level. Environmental Order Plans can be included into high physical plan level. Master Development Plan level and Implementation Plan levels are characterized by the local physical plan levels.

The plan-making process represents the comprehensive rational planning process that is divided into the four main groups, these are listed in a hierarchical manner as follows:

a. Spatial Strategy Planning

b. Environmental Order Plans

c. Master Development Plans

d. Implementation Plans

Basic characteristics of the spatial planning system can be itemized as follows:

– All plans have to obey the decisions of the current upper plans and all of them have to direct the lower plan hierarchies.

– Regarding land use and built-up environment, development authorities and the all stakeholders in the society and citizens have to obey the decisions of Spatial Strategic Plans, Environmental Order Plans, and Master Plans.

– All other plans have to maintain inputs for these plans.

– Local governments have to obey the decisions of Environmental Order Plans while preparing the Master Development Plans and Implementation Plans.

The main characteristics of these plans have to be identified to understand the convenience of the system for sustainable urban planning and sustainable urban development.

5.1. Spatial strategic plans

These plans are appreciated as the level which integrates the national development policies and regional development strategies with the spatial level. In other words, this plan level stays between the socioeconomic level and the physical plan level. It helps to transform the decisions of national and regional plans into spatial planning of the localities.

It has the following several characteristics:

– Relates the National Development Policies and Regional Development Policies with the spatial level.

– Defines the spatial strategies related to the urban growth, transportation system, and social and technical infrastructure.

– Relates the spatial strategies with sectoral strategies.

– Especially prepared by using schematic and thematic graphic language on 1/250,000 or 1/500,000 or upper scales.

– Prepared for the regions or localities that are appreciated as this plan is necessary.

Planning principles for Spatial Strategic Planning are as follows:

– Conservation of historical and cultural values;

– Minimizing the threats and risks of disasters;

– Maintain sustainable use of resources;

– Distributing infrastructure, service, and production facilities suitable for development policies along urban and rural areas ;

– Maintain public utilization and efficiency in resource use, productivity, and transparency;

– Maintain multidisciplinary mode of planning with contribution of institutions that affect or can be affected from planning such as Development Agencies, Non-Governmental Organizations (NGO's), Chambers, Local Governments, Universities, Private Sector Represents;

– Maintain spatial harmony;

– Making of necessary spatial arrangements to establish innovative, flexible, and competitive economic structure;

– Adaptation capability to the changing conditions.

Elements of the Spatial Strategic Plan are as follows:

– Sectoral and thematic decision sheets/maps on settlement systems, transportation, water, risky situations, infrastructure, economy, and specialized regions.

– Sectoral or thematic sheets/maps of several regions if it is prepared for a region.

– Spatial strategy report including vision and priorities, principles, strategies, sectoral decisions, provisions of the plan, and main headings of the action plan.

Application and monitoring of the Strategic Spatial Planning is maintained by the activity reports of the relevant institutions. These are evaluated in accordance with these institutions and revised if necessary.

As it is the one of the most important stages of all plan levels, also quite a few information and data are collected. These data are with regard to dangerous areas, all types of ecosystems, water resources, site areas, settlement systems, national parks, demographic data, sectoral data, sectoral plans, national development plans, transportation networks, housing strategies, etc.

5.2. Environmental Order Plans

These plans are implemented at least on the province level or especially at statistical regional units that appreciated as the high-level spatial plan performed at the scale of 1/100,000 or 1/50,000.

Planning principles and essentials can be listed as follows:

– It has to be coherent with spatial strategic plans;

– Regional dynamics and potentials have to be evaluated;

– Sectoral investment decisions that can affect spatial decisions of the related public institutions have to be evaluated;

– Appropriate ecological and economic decisions have to be decided together to the sustainable development;

– Conservation and improvement of the natural structure and the landscape such as historical and cultural assets, forests, agricultural lands, and water resources;

– Maintain unity of land use to preserve continuity of ecological balance and ecosystems;

– Generate routes of transportation network;

– Determine preventing strategies and policies for the environmental problems and generate land use decisions;

– Take precautions for the destructive effects of disasters.

Within the process of preparing Environmental Order Plans, these types of analyses have to be fulfilled:

– All types of borders (administrative, planning etc.);

– Administrative and regional structure;

– Physical and natural structure;

– Site areas and other conservation areas, and sensitive areas that have to be preserved;

– Economic structure;

– Sectoral developments and employment;

– Demographic and social structure;

– Urban and rural settlements and land use;

– Infrastructural systems;

– Open spaces and green areas;

– Transportation systems;

– Risky areas;

– Military areas and security zones;

– Public projects and investment decisions regarding the planning area;

– Hydrological and hydrogeological areas;

– Environmental areas and affected areas.

Environmental Order Plans are prepared at the scales of 1/100,000 or 1/50,000. On the other hand, the sensitivity of the 1/25,000 scale is expected. All necessary information and documents are gathered by experts and land works. Geographic information systems and remote sensing methods have to be applied in the planning process and a database is generally structured after the work.

If the Environmental Order Plan is not satisfied with the needs or the vision, aims, targets, strategies, and policies are not satisfied with the upper decisions, then a revision of the whole plan has to be made. The revision becomes necessary at the following special circumstances:

– When the population needs are not satisfied,

– When new regional investment opportunities arise which change the basic strategies and policies of the plan,

– When the new land use demand is occurred that has a potential effect on regional issues,

– When there are changes at regional dynamics through new developments.

According to these factors, a plan change can be made, which cannot damage the integrity of the plan decisions and plan continuum.

5.3. Land Development Plans

This plan is considered as the special spatial plan that allows the application of Spatial Strategic Plan decisions and Environmental Order Plan decisions at the settlement level. These urban plans include constituted alternative solutions for settlement patterns and their growth trends. Land use decisions are the main part of Development Plans. In addition, there are decisions on conservation, limitation, organization, and application.

Land Development Plans are divided into two parts as Master Development Plans and Implementation Plans. All types of Land Development Plans are prepared and approved by the related municipalities.

These plans have to obey the decisions of upper scale plans. In a plan-making process, geological and geotechnical land etudes have to be completed, field surveys, threshold analysis, and other compulsory analysis have to be accomplished, and opinions of the public institutions have to be taken into consideration.

There are several principles that dominate the preparation and application of Land Development Plans in general. These can be listed as follows [11]:

a. Obey the state of law: Plan has to depend on legal basis.

b. Principle of clarity: Urban planning decisions and plan have to be known by every stakeholder in the public. This can be maintained by participation of different institutions and citizens in the planning process.

c. Principle of universality: Especially, Master Development Plans and upper scale plans show the general and common decisions, and they do not deal with details.

d. Hierarchy principle: There is a hierarchical relation between different plans. The main aim is coordination between plans. Upper scale plans have to direct lower scale plans. Their decisions have to be coherent.

e. Public welfare: One of the important aims of making Land Development Plans is maintaining public welfare. Primarily, urban spatial plans are prepared for public welfare.

f. Obligatory characteristics: Plan decisions have to be obeyed by every stakeholder in the society. Also, coherence of various level decisions is important.

g. Flexibility: Urban plans have not to be changed without very important reasons. On the other hand, due to dynamic social structure, and rapidly increasing population, these plans have to be changed within a given time period. So, flexibility covers revision plans or additional plans.

h. Long range characteristics: Land Development Plans are long range plans. New regulation defines the period as 20 years.

i. Being scientific: Scientific data and information have to be collected and evaluated in the planning process. All analyses that have to be carried out are related to this principle.

j. Participation principle: That means individuals taking the decisions about the physical and built-up environment and collaboration in the planning process. According to the Regulation of Making Spatial Plans, participation tools include surveys, opinion research, meetings, workshops, information services, etc. It becomes important to take decisions about all stakeholders.

5.3.1. Master Development Plans

Master Development Plan is prepared at the scale of 1/5000. Plan determines the alternative growth and development structure of the following:

– All kinds of administrative, planning boundaries;

– Area restrictions for special conditions;

– Housing areas and housing development areas;

– Administrative centers;

– Trade and work centers;

– Industrial areas and warehouses;

– Tourism areas;

– Site areas;

– Conservation and resource areas;

– Energy transmission lines;

– Open and green spaces;

– Social reinforcement areas, such as, education, health, and culture;

– Technical infrastructure such as roads, auto parks, different transportation modes, water infrastructure, waste treatment, and garbage areas.

While preparing Master Development Plans, the data listed below are analyzed and land work studies are conducted:

– Administrative boundaries;

– Geological, geomorphological, and hydrogeological structure;

– Characteristics of different settlement areas and spatial growth tendencies;

– Renovation and transformation zoning;

– Climate;

– Vegetation;

– Soil structure and agricultural use of the land;

– Flora and fauna (ecological searches) ;

– Site areas, national parks, natural parks, wildlife conservation area, and protected water basins;

– Forests, pasture lands, and highlands;

– Cultural and touristic conservation and development regions;

– Industrial areas and organized industrial sites,

– Landscape elements;

– Demography;

– Social and economic structure;

– Transportation systems;

– Environmental problems;

– Logistic center areas;

– Sectoral structure (agriculture, industry, services, transportation, energy, mining, and construction) ;

– Solid waste and recycling facilities;

– Waste water discharge areas and facilities;

– Mining areas;

– Military zones;

– Risky areas and risk plan if exist;

– Decisions of environmental order plans.

The Master Development Plan process consists of the following six steps:

i. Determination of the planning area;

ii. Gathering planning data:

– Opinions of public institutions;

– Field studies:

a. Building characteristics (storey heights, use, type, etc.);

b. Survey study (housing, working places, industry, administrative units, social utilities, etc.);

iii. Analysis and synthesis: After field surveys and evaluation of various opinions, some sorts of analyses have to be completed. These data are superposed and then synthesis maps are produced. Hence, it is able to determine appropriate land for settlement growth. Analyses are executed on land use, slope, natural factors, geology, physical data on buildings, population density, structure analysis, land ownership pattern, land values, transportation networks, transportation zones, and upper scale plan decisions. Synthesis studies are conducted on natural threshold, and appropriate zones for settlement.

iv. Clarify plan decisions: Different plan alternatives are produced after the analysis part and evaluation of the socioeconomic structure. These alternatives are evaluated and planning decisions are made.

v. Plan drawing and report preparation: Once the best alternative has been chosen, the plan is nearly completed. Urban Master Development Plan is a unity with its drawings and plan report.

vi. Evaluation and approval of the plans by Municipality Councils.

5.3.2. Implementation Plans

This stage of the spatial planning system represents the application of all planning decisions on the urban space. It is prepared at the lot and parcel level with the scale of 1/1000. That is why it is called as "Implementation Plan." It can be performed at once as well as stage by stage. It is the plan level, which urban standards have applied at the lots level. Building conditions are the main point of Implementation Plans. Especially, it is the best plan type at planning level to take decisions at third dimension apart from the urban design stage.

Some important principles of Implementation Plans are as follow:

– Defining building conditions, such as, lot coverage ratio, floor area ratio, floor area index, building heights, setback distances, etc.;

– It is important to develop design principles for maintaining accessibility of handicapped, elder ones, and children to all social infrastructure and urban uses;

– It is essential to develop pedestrian and bicycle networks and bicycle parks;

– Natural and historical site areas are given importance within the plan.

As similar to upper planning scales, some set of data have to be gathered while preparing the Implementation Plan. In light of these data, the following analysis and research have to be performed:

– Analysis of the decisions of Master Development Plan,

– Boundaries of the planning area,

– Existing building density,

– Location of building lots and characteristics,

– Construction and setback distances,

– Existing population density and population distribution,

– Social infrastructure facilities,

– Technical infrastructure facilities,

– Land ownership pattern,

– Site areas and officially registered assets,

– Accessibility to services,

– Geological etude works,

– Topography and thresholds,

– Water resources,

– Transportation networks and nodes,

– Open space building relations,

– Assembly areas,

– Auto parks and capacities,

– Pedestrian and bicycle networks,

– Service areas and locational choice characteristics,

– Lands of urban design project areas,

– Logistics areas.

Implementation Plans contain notations of the following areas or functions:

– All types of boundaries including administrative, planning and restriction zones, and risky area zones;

– Housing areas;

– Social and cultural areas;

– Trade areas;

– Industrial areas;

– Recreational and green areas;

– Cemeteries;

– Special areas like military zones.

Adjustment tools utilized within plan-making process are zoning, surveillance of building lots, and norms and standards. Zoning consists of functional, density, and height zoning. Division of building lots is needed for the implementation of architecture of buildings on building parcels. Standards are used for the comfort, livability, and sustainable living of urban residents.

Implementation plans are implemented on urban space with programs. These programs are especially prepared for the five-year period. It has to be executed successively. Continuity is the essential principle in this process. Municipalities have to prepare programs within a time period of three months after the completion of the plan.

6. Criticism of the Turkish planning process with regard to sustainability necessities

Besides the structural shortcomings of the Turkish planning system, emphasis is on the critics of the Turkish planning process and planning hierarchy with respect to sustainability criteria. Even in the Habitat-III National Report of Turkey, it is claimed that sustainable urban planning in Turkey takes its reference from the physical dimension. The report claimed that there are shortcomings in proper handling of social, cultural, economic, and ecologic dimensions of planning [12]. Also, there is no evidence on how relations between economy–ecology and society are shaped within the planning process. On the other hand, the main bottleneck of the Turkish planning system is defined as follows [13]:

- Lack of a spatial planning system integrated with national development planning,

- Lack of a cooperative planning vision and strategy,
- Partial implementations in planning and various conflicts between decisions of different public authorities,
- Numerous authorities for the same spatial scale,
- Lack of coordination between institutions.

With reference to the Turkish planning process explained in the former part, there are some shortcomings of the planning process related to sustainability essentials.

First, there is a one-way flow of decision-making takes place in the system of Turkish urban planning. An elitist type of planning is dominant. Spatial decisions of all scales are taken at bureaucratic levels and they are dictated to citizens by the plan itself. In this elitist type of planning, planner's role is taken only at technical level. This causes a problem to the rant-oriented decision making of municipalities as they are the approval mechanism of the Master and Implementation Plans. That is, especially, not the problem of planning process but it is entirely an ethical problem. These types of actions commonly bring out many legal problems.

The Turkish urban planning process also suffers from the absence of feedback mechanism. Feedback is executed as preparing ``revision plans" and ``local plan changes." This situation really leads to the waste of resources because all plans need an important amount of effort. The lack of feedback and elitist type of decision making leads to legal problems that are always tried to be solved at administrative courts. Every case in the courts makes plans impermanent and ineffective and leads to waste of resources.

Another important problem is the lack of a monitoring process in the Turkish urban planning system. As it is the case of the lack of feedback, the lack of a monitoring process is also tried to be covered by revision plans and local plan changes. The lack of a monitoring process is an obstacle on the control of the Master and Implementation Plans. For this reason, it is compulsory to prepare a Master and Implementation Plan for all of the settlements, most of the settlements in Turkey are perceived as a nonplanned locality. Decisions of original plans cannot find a chance to be applied till the deadline of the plans but are always changed by the revision plan or plan changes. There are definite development programs for the implementation of the urban plans but this mechanism is not organized as a monitoring process. However, these programs do not get a chance to be implemented as the construction activities are totally in the hand of speculator contractors.

The sustainability concept is emphasized in the last Regulation of Plan Making Process, as it was described in former parts. On the other hand, there is no evidence that how sustainable urban development can be achieved. Especially, sustainability indicators are not included into planning process, even no such indicators exist in the Turkish planning process. A lot of necessary or unnecessary data are collected throughout the planning process but neither of them are used for evaluating the sustainability indicators. The data collected at higher levels cannot be used at lower scales. With the collected data, it will be easy for evaluating sustainability indicators if they exist within the plan process.

One of the most important bottlenecks of the Turkish planning process is the lack of participation even though the "participation" concept takes place in the Turkish planning system. On the other hand, this does not fulfill the conditions that sustainable type of planning requires. In Turkey, the participation of citizens to planning is achieved by two ways. One is participation in questionnaires and surveys before the plan-making process and the second one is related to gathering information from the prepared plan itself. After completion of Master and Implementation plans, these plans are exhibited on municipality boards. This is for gathering information of the landowners about their parcels. Citizens can make their objections to the plan within a month time. Usually, nobody has information that plans are exhibited on the Municipal Boards unless they see or heard about by chance. These two situations about participation have no relation with participatory democratic planning. As a result, citizens are given no right to have opinions about the plan throughout the plan-making process. Even, they did not know the existence of such a planning process. This condition is also valid for other public institutions. The court cases of the plan conflict between the Public Treasury Office as owner of public lands and Municipalities as making and approving urban plans, which are the best examples for this situation.

Though there is an obligation to take decisions about all institutions in the planning process, this is not a guarantee that the participation of institutions to the planning process is treated in a proper way. These decisions are taken from only relevant institutions on relevant lands, not for the whole plan.

All these findings clearly indicate that the Turkish current planning system is not emphasizing governance though this concept is given much importance in the stage of Strategic Spatial Plans. As the hierarchy of the plan scales is lowered, government-dictated decisions become dominant instead of governance principles.

7. Conclusion and recommendations

There are increasing debates on the subject whether the rational comprehensive method of urban planning can be able to ensure sustainable urban development or not, and whether this type of planning approach has the requisites of the sustainable urban planning. The rational comprehensive planning model is a pure top-down hierarchical approach using deductive point of view. The rational comprehensive type of planning is dominant in many countries, as it is the case in Turkey. This situation in Turkey depends on the law accepted in 1985. However, there were many efforts to change the planning system, it continued as a dominant planning approach with minor changes and adjustments in passing periods.

The sustainability concept and sustainable urban development require many new intentions for the planning or the plan-making process. Governance, participation especially in participatory democratic planning, use of indicators, existence of efficient feedback mechanism, and monitoring process are some novelties for the sustainable-based urban planning.

Regarding the existing planning approach and the sustainable urban planning concept, the most important question is how to adapt the sustainability concept to the rational comprehensive planning process without a paradigm shift in the Turkish spatial planning system.

As mentioned earlier, the Turkish planning system is a rational comprehensive process having substantial efforts to maintain sustainability. With regard to deficiencies of the planning process, several measures or precautions can be taken within the existing plan-making process in order to adapt sustainable planning requisites.

This can be held without a total system change, but with some serious and radical adjustments in the process. National and regional decisions can be made from the upper scale decisions but these decisions have to be transformed and adapted to the levels of Master Development Plans and Implementation Plans. Most problems about urban sustainability arise at these lower scales.

First keyword for the change is the participation. Citizen participation in the process has to be achieved in an active way. Meetings and open workshops will be the main instruments. People who own land in the planning area can be informed completely by this way and as this situation is legalized decisions will be taken at consensus and there will be no need for court cases. Planners have to persuade these people in a peaceful manner and this is possible when face-to-face active participation takes place in the planning process. The situation does not affect the role of urban planner as a decision maker, but adds a new mission and a new role for planners as a persuader and intermediator. Though the planning process can take a bit longer, implementation will be guaranteed. There will also be several feedback on decision and all these feedback lead to healthy decision making.

The second important factor is the monitoring process. Monitoring can be completed easily with the help of participation. Here, participation is related to the acts of several public institutions and landowners that control and direct the land development programs.

Monitoring processes are mostly used for the control of the sustainability indicators that have to be included into the planning process.

The sustainability indicators can be added to the process at different plan levels such as at the Environmental Order Plan level, Master Development Plan level, and at the Implementation Plan level. These indicators can be related to the following sustainability criteria:

At Environmental Order Plan Level (scale of 1/50,000)

- Conservation of natural and cultural heritage,
- Improvement of transportation,
- Resource saving (reuse of technologies and recycling),
- Conservation of agricultural areas,
- Qualify services,
- Satisfying all actors,

- Strengthening local economy,
- Generating employment opportunities for the local economy,
- Maintain infrastructure standards,
- Fresh water supply,
- Waste disposal.

At Master Development Plan and Implementation Plan levels (scale of 1/5000 and 1/1000)

- Mass transit opportunities,
- Less noise pollution,
- Protection of cultural and natural assets,
- Reusing,
- Public spaces,
- Contaminant minimization,
- Sanitation facilities,
- Open areas,
- Improvement of living conditions,
- Energy and water saving,
- Accessibility,
- Establishment of effective infrastructure,
- Livable environments and neighborhoods,
- Qualification and adequacy of services
- Interaction with nearby settlements
- Energy saving
- Adaption of local design styles

With several interventions to the Turkish planning process, it has a potential to adapt sustainable type of urban planning. These interventions are related to the variables of citizen participation, monitoring, and sustainability indicators as these are the most common and important determinants of sustainable urban planning. Sustainable urban planning is not a new paradigm. It is a broader phenomenon that all types of planning approaches and planning processes have to adapt.

Author details

Okan Murat Dede

Address all correspondence to: okandede@gmail.com

Amasya University, Amasya, Turkey

References

[1] Naess, P. Urban planning and sustainable development. Eur Plan Stud 2001;9(4): 503–523. Taylor and Francis Publishment, ISSN: 0965-4313 (Print).

[2] Yazar, KH, Dede, OM. Sustainable urban plannıng in developed countrıes: lessons for Turkey. Int J Sus Dev Plan 2012; 7(1): 26–47. DOI: 10.2495/SDP-V7-N1-26-47.

[3] Tsenkova, S. Sustainable Urban Development in Europe: Myth or Reality, Events and Debates. Blackwell Publishers Ltd., USA, 1999.

[4] Atay, İ. Sustaınability Measurement in Urban Planning Practice: Evaluating the Environment Plans of the Cities in Aegean Region. Master Thesis. Graduate School of Engineering and Science of İzmir Institute of Technology, July 2009.

[5] UN & Habitat: Inclusive and Sustainable Urban Planning, A Guide for Municipalities. United Nations Human Settlements Programme, Volume 1: An Introduction to Urban Strategic Planning, Dec, 2007. ISBN Number (Volume): 978-92-1-131929-3.

[6] Rosales, N. Towards a design of sustainable cities: incorporating sustainability indicators in urban planning. In: 46th IsoCarp Congress, Nairobi, Kenya, 2010.

[7] Yazar, KH. Sustaınability Measurement in Urban Planning Practice. Doctorate Thesis. Ankara Üniversitesi/Sosyal Bilimler Enstitüsü, 2006, 306 pp.

[8] Carriou, C, Ratouis, O. Is There a Model for Sustainable Urban Planning? [Internet] 2014. Available from http://www.metropolitiques.eu/IMG/pdf/met-carriou-ratouis-en.pdf [accessed 2015-12-19].

[9] Tarlak, Ö. İmar Yasalarının Gelişimi Üzerine Düşünceler. 2002, Çağdaş Yerel Yönetimler Dergisi, cilt 11 sayı:3.

[10] Mekansal Planlar Yapım Yönetmeliği [Internet]. December 2014. Available from http://www.mevzuat.gov.tr/Metin.Aspx?MevzuatKod=7.5.19788&MevzuatIliski=0&sourceXmlSearch=mekan [Accessed 2016-02-03].

[11] Kalabalık, Halil. İmar Hukuku Dersleri, Seçkin Yayıncılık, 2015. 735 pp. ISBN: 978-975-02-3407-1.

[12] Habitat-III Turkey National Report [Internet]. December 2014. Available from http://unhabitat.org/wp-content/uploads/2014/07/Turkey-national-report.pdf [accessed 2015-12-10].

[13] Dede, OM, Ayten, M. The role of spatial planning for sustainable tourism development: a theoretical model for Turkey. Tourism, 2012; 60(4): 431–445. UDC: 338.484:502.131.1]:71(560).

6

The Environmental Dimension of Urban Design

Ilaria Giovagnorio and Giovanni M. Chiri

Abstract

International research on sustainable architecture ascertained the responsibilities of urban forms for buildings' energy-environmental performances, highlighting the necessity to broaden the field of intervention in urban design. Furthermore, goals concerning the sustainable city increased design complexity, due to the involvement of different interrelated disciplines, which modified design processes by incorporating external contributions. In particular, environmental analyses are growing in importance and need to be reintegrated into the urban project at the conceptual stage. This 'environmental awareness' accompanies the history of the city, and numerous pieces of evidence clearly show the mutual and in-depth relationship between urban form and local microclimates. Lessons from the ancients constituted the fundamentals in urban design until the Modern Movement, during which knowledge of the past also influenced the work of G. Vinaccia, an Italian pioneer in microclimatic urban design. After the World War II, most of the lessons had been forgotten in favour of technology systems that have since revealed their failures. The current design condition requires a discovery of past abilities, coupling them with contemporary scientific advances. This work introduces a methodology through which to integrate current urban design processes with environmental data and analyses. It is illustrated through a case study and is supported by software.

Keywords: urban design, environmental design, sustainability, sustainable architecture, Gaetano Vinaccia

1. Introduction

During the last few years, actions directed towards the sustainability of architecture mainly involved buildings. Nevertheless, several international research groups have ascertained the influence of the urban form on local microclimate.[1] In fact, the way in which buildings are arranged on a site heavily affects their owners' energy performances and environmental behaviours. The growing awareness of these issues is gradually attracting the attention of both international organizations and international researchers towards the urban dimension (urban block or district), thus broadening the field of architectural interventions. In this light, urban design, which lies in-between urban planning and architectural design, has become the appropriate tool through which to operate in achieving sustainability's goals. This recommendation, however, is not novel. In 1976, the European Commission suggested it during the first United Nations Conference on Human Settlements, 'Habitat' (Vancouver), in which the 'design of human settlement' was recognized as a strategic matter in contrasting the «social, economical, ecological and environmental deterioration» of urban areas [1]. Nevertheless, the latest 'sustainable urban design' seems to increase its own complexity due to the involvement of different interrelated disciplines, such as fluid dynamics, climatology, technical physics and computer engineering. This interdisciplinary perspective demands updates to the design process in order to incorporate external contributions while maintaining architecture at the heart of the process. In particular, both environmental data and analyses need to be reintegrated into the urban project from the design's initial phases. In fact, this has been an ordinary consideration in the history of architecture, in which several pieces of historical evidence prove in a clear way the in-depth relationship between urban form and local climatic conditions. Examples occur throughout history, starting from vernacular settlements until the Modern Movement's masters, and most of the 'environmental' knowledge of past designers is also stated in important recommendations contained in ancient manuscripts (Vitruvius, Aristotle, Varrone, Columella, Palladio, etc.). Nevertheless, most of the established abilities have been forgotten in favour of building technology systems, with an illusory belief in their supremacy. The energy and environmental crises of the 1970s clearly displayed their failure, encouraging designers to rediscover projects' environmental components as a function of a sustainable future. In this light, the past's lessons on 'sustainable practices' have grown in importance, especially when coupled with scientific and informatic progress. Computer advances, in fact, make it possible to study increasingly broad areas (CitySim, ENVI-met, Virtual Environment) and contribute to supporting urban design's processes with environmental analysis software. The latter acts as useful 'feedback' tools, able to verify (qualitatively) the environmental behaviour of the project's concept by taking into account different climatic data (air temperature, relative humidity, wind velocity and main directions, etc.) and form parameters (sky view factor, aspect ratio). In this way, designers experiment with a new methodology in the project's development, one that allows them to evaluate the initial urban proposal and, gradually, to modify it in relation to the main criticalities that emerged from previous environmental analyses' results. Nevertheless, modifications must respect a project's fundamentals and collaborate in the improvement of its overall performance.

---

[1] Examples are the Martin Centre for Architectural and Urban Studies of Cambridge; the 'Urban Physics' research group (ETH, TU Delft and Cypro University); the Centre for Advance Urbanism and the SENSEable Lab of MIT; the Harvard Centre for Green Buildings and Cities and the City Form Lab; the EPA's EPFL Solar Energy and Building Physics Laboratory (Luisanne); the Environment People and Design (ePad) Research Group in Nottingham (UK); the Berkeley's Center for Environmental Design Research (U.S.A.) and the UCL-Institute for Environmental Design and Engineering in London (UK).

In this light, this work aims at introducing a methodology that is able to integrate current urban design processes with environmental data and analyses. The method is illustrated through an urban-design case study in Sardinia (Italy). It concerns the urban expansion of Monserrato in developing a new university district of 1.4 km.[2] All the design phases were supported by Heliodon and ENVI-met software. The work was preceded by a study on the main directions of current 'sustainable design', highlighting both the necessity (and the advantages) of broadening the field of intervention to include the urban dimension (urban blocks or districts), and its current interdisciplinary character. Previous goals are also supported by several examples in the history of architecture, which show the ancient roots of this environmental approach.

2. Sustainable design: from buildings to the city

As stated above, the relationship between a building and its context has a heavy influence on both local microclimatic conditions and a building's energy-environmental impact, shifting designers' and planners' attention from buildings to the block dimension. The urban structure substantially affects a building's access to sunlight and ventilation, defining its capability to exploit passively the heat produced by solar radiation on vertical surfaces, the natural daylight and ventilation of inner spaces, as well as the reduction in environmental and noise pollution. The necessity of this 'change of scale' has been affirmed, in the last few years, in several national and international scientific works, in which authors highlighted the achievable benefits both in relation to climate change goals and energy issues. Among the most significant attestations, the U.S. Green Buildings Council stated that the neighbourhood level was «the primary scale at which building professionals can begin to address impacts of climate change». In fact, «the design and pattern of neighbourhoods [...] play an important role in amplifying or dampening climate effects» [2]. Evidence of this is the Urban Heat Island phenomenon (UHI), fostered by cities' higher densities and the *albedo* properties of common urban materials, and the more frequent extreme weather events, such as flooding, strengthened by urban pavements' waterproofing and inadequate storm-water runoff.

Furthermore, in relation to the energy issues, several pieces of national and international research supported the block dimension as the only possible way to decrease buildings' consumption and, therefore, to improve their environmental impacts. A recent study by Hachem, Athienitis and Fazio [3] showed how different layouts of the same building typology (terraced houses) modify their energy consumption, increasing their values between +6 to 8 and +25%. Orientation and buildings' arrangements are considered central in the project's process by its authors, conferring on the initial phase of the design huge energy responsibilities, which were estimated as being between 65 and 85% of successive building energy demands.

[2] The International Congress of Modern Architecture focused on the work of Hannes Meyer, Mart Stam, Walter Gropius, Erns May, and Alexander Klein on *Existenzminimum* ("minimum dwelling"), aimed at reducing building costs (including the cost of land consumption) by a reduction of the worthless surfaces of houses. Research on typology reached some standard formulations of the new working-class building environment, in which the amounts of daylight, fresh air, heat, and silence were radically maximized.

The results of this study are comprehensible in relation to the microclimatic consequences of different buildings' arrangements. The influence of the modification of urban geometry on the urban microclimate was well known until the 1970s, in the main works of the climatologists T.R. Oke [4] and H. Langsberg [5], which became the theoretical basis of reference for most of the successive work on environmental design. As stated by Givoni in 1989 [6], and reported later in Steemers' and Ratti's work [7], the influence of climatic change on outdoor temperature, solar radiation and wind speed, caused «by 'the structure' of the city [...] result in modified energy consumption». An urban grid's orientation, density and a building's typology heavily influence the wind pattern and the solar access in urban areas, while also defining the microclimatic conditions of outdoor and inner spaces. Further significant studies on the energy and environmental 'costs' of urban form have been developed during the last few years by several scholars (K. Steemers [7–10]; V. Gupta [11, 12], R. Compagnon [13], R. Knowles [14], E. Ng [15, 16], F. Allard [17, 18] and B. Blocken's research group [19, 20], including J. Carmeliet and M.K.A. Neophytou [21]).

Interest in this 'change of scale' in sustainable design is slowly extending from scientific research into concrete actions. In 2010, the European Commission introduced the *Advisory Group ICT Infrastructure for Energy-Efficient Buildings and Neighbourhoods for Carbon-Neutral Cities* [22] and, in June 2014, the European Urban Knowledge Network (EUKN), together with the Cellule Nationale d'Information pour la Politique Urbaine (CIPU), presented the *Certification Systems for Sustainable Urban Neighbourhoods* «as tools to evaluate and maintain high quality standards in urban development» [23]. Although practical initiatives are still few in number, they clearly envisage architects and planners' future goals.

3. Environmental lessons from history

Even though the application of sustainable environmental theories and practical solutions in the urban dimension appears to be novel today, it was actually a common praxis in the past. It concerned both vernacular settlements and the foundation of ancient cities, such as Greek and Roman examples. Furthermore, the ability of past designers to plan in accordance with local weather and environmental features guided their constructive practices for the following centuries. Palladio's works included important environmental devices, although these were mainly applied to individual villas, whereas the Hygiene Movement and the successive Modern Movement (XIX-XX centuries) extended them to the urban dimension.

3.1. Vernacular settlements

According to G.D. Carlos et.al. [24], the morphological consolidation of vernacular settlements is constrained by three main concerns, among which are local microclimatic conditions. A building's aggregation system is «the primary instrument to attenuate the relation between the inhabitants and the natural elements», controlling a building's exposure, both to the sun and the wind. Climatic challenges, together with available construction technologies and relief constraints, usually originated «a *diffuse* or *compact* spatial organization, and a *scattered* or *dense* building aggregations». Furthermore, the energy supply, based mainly on wood and animals' muscle strength, contributed in developing a 'discrete' urban model, limited in its growth by the presence of natural resources in the surrounding area [25]. Examples exist in every continent, and each morphological configuration responds through its own buildings' spatial organization to the local conditions. A significant example of hot-dry climate architecture is the medieval citadel of Shali in the Oasis of Siwa (Egypt), as described by A. Picone [26]. Here, housing aggregation creates an absolutely compact nucleus that seems from the outside to be «one single large building» [27] (**Figure 1**). This compactness allows the reduction of surfaces facing solar radiation, thus increasing buildings' mutual shadows. Since the traditional Arabian courtyard is missing in Siwa's houses, it is the streets (very small with curved geometries) that perform as temperature regulators.

Figure 1. The Oasis of Siwa (Egypt), *«one single large building»*.

A second example of a vernacular climate-sensitive architecture is the 'solar communities' of Pueblo Indian tribes of south-west America, built during the eleventh and twelfth centuries A.D. The most well-known ruins of the Long House at Mesa Verde, Pueblo Bonito and Acoma Pueblo (or 'sky city') in New Mexico prove that their sophisticated design skills were able to provide year-round comfort for people dwelling within these structures. A study by R. Knowles [28], conducted during the 1980s, deepened knowledge of the settlements' environmental behaviour, focusing on their correlation with solar dynamics. Concerning the Long House case study, both the location and the siting of structures within the cave collaborate to «mitigate extreme environmental temperature variations by responding to the differential impact of the sun during summer and winter, night and day» [28]. Both the southern orientation and the conformation of the cave allowed the Ancestral Puebloans to exploit passively the low winter sunrays to heat houses and, at the same time, to shade them from the high summer sun. Building's adaptation to solar dynamics also characterized Pueblo Bonito. Furthermore, in this case, the dwelling «relied primarily on its form to respond to solar access». Its semicircular form, coupled with the buildings' disposition to take the form of terraces descending towards the city centre, helps to mitigate the extreme seasonal climatic effects. The use of terraces is also common in Acoma Pueblo buildings [29]. Here, the urban structure is organized in three main rows oriented east–west and each row is distanced from the others in order to ensure the exposure of the entire facade to winter rays. Openings' orientations and dimensions, the different employment of natural materials and the buildings' shapes contribute effectively to their overall performance. In particular, «even more surprisingly is the orderly town plan that guarantees all residents full, equal access to the sun's heat» [30].

A final example is the vernacular underground settlements and their bioclimatic advantages. An interesting research project conducted by Vegas, Mileto, Cristini, and Checa in 2014 [31] collected several examples of cave dwellings, highlighting their environmental weaknesses (land consumption, lack of ventilation and the risk of dampness) and potentials (great insulation potential, great heat retainer and very low energy costs). Among the outstanding cases, the Berber dwelling of Matmata in Tunisia [32] responds to the extreme conditions structuring its hypogean houses around a circular courtyard, located at approximately 11 m below ground level. This has a 5–12 m diameter, and it is provided with a tank for the collection of rainwater. The circular form allows part of the courtyard to always be shaded, whereas the underground condition protects it from hot sandy winds.

3.2. The 'oriented' urban grid: the new city

According to Secchi, «the grid is the ordering element, par excellence» [33]. Its application dates back centuries and «certainly, it is not a founding shape of the Occident culture». Important examples exist worldwide and all of them share similar organizing principles: an urban grid, oriented along the cardinal points that define the dimension and orientation of blocks, and a building typology, which was mostly southern oriented and provided by transition components, which mitigated the outer climatic conditions. Repetition of such communal features is probably due to similar environmental conditions, since, according to Los's research [34], the first ancient 'solar cities' originated between the 30th and 40th parallels; thus, they were characterized by hot and temperate climates. Among the oldest examples El Lahun working-class neighbourhood (Egipt, 1800 B.C.), Babylon (Mesopotamia, 2300 BC) and the Pakistani settlement of Mohenjo Daro (Indus civilization, 2750 BC) where «street grid [was] oriented according to prevailing winds» [35].

The grid plan shows so basic a principle that it is not necessarily linked to influencing or dependent relationships among populations. Furthermore, several scholars often find in its recurrence the evidence of cultural exchanges among Western and Eastern communities. As stated by Castagnoli [36], the use of an urban grid defines, upstream, the existence of the authorities' planning intentions; and for this reason, it is a common feature of foundation cities (as colonies, military camps and reconstructions). Although the application of a grid structure can be ascribed to different purposes (political, religious, military, etc.), most scholars agree on both the importance of its orientation and how the orientation is strongly related to astronomical issues: in particular, to the sun's path and the winds' main directions.

Although the grid's origin is not precisely known, nor is the mutual 'contamination' of different cultures, the first use of a north–south-oriented urban grid is usually attributed to Hippodamus of Miletus. Furthermore, it is important to point out that «it can be proposed that the name of Hippodamus goes not connected with the far ancient orthogonal system, but with that developed plan for a regular reticulated city, which can be identified during the fifth century [...] therefore [Hippodamus was] not an icon, nor a simple theoretician, but an urban planner, who played an important role in the urbanism of the fifth century». This kind of planning, which could not be documented before these data, is characterized by two main features: the presence of a complete urban plan and an in-depth study of residential blocks. Von Gerkan insists on the topographical and sanitary meaning of Greeks cities' orientations, whose purposes were more suitable wind exposure and access to the sun's rays [37]. Both Hippodamus's works and the later essay by Hippocrates, *On Airs, Waters and Places*, attempted to reconcile urban studies with environmental theories, paying attention to the city's healthiness. In reference to this, it is interesting to underline that Hippodamus was also called a 'meteorologist' (meterewlogoς) by Hesychius and Photios.

Ruins of the Greeks' 'solar cities' show as urban grids, and blocks were designed to take into account the different seasonal heights of the sun in order to guarantee the correct solar access in buildings over the whole year. Priene and Olynthus's plan testified to these 'solar principles': in particular, Olynthus's block-long row of houses distinguished the northern façade (thicker walls without openings) from the southern one, which was open to the sun. In addition, both the houses' and the rows' distance from one another contributed to the 'solar' strategy, avoiding the buildings' mutual shadows in wintertime. The case of Delus, moreover, confirms the importance of the southern orientation, which was declined to single houses where topographic conditions impeded the application of the wider urban grid. The real functioning of the ancient Greeks' environmental urban design was finally proved by Tatcher's studies on rooms' solar-heating capabilities, whose results demonstrated comfortable performances «on 67% of the days during the colder months» [38].

Even if most Roman cities were based on the same reticular structure, their models appeared to be less strict, showing several rule variations. Furthermore, the grid's orientation to the functions of the sun and the winds still remains a feature that it has in common with the Greek model.

According to Vinaccia's statements [39] (Italian architect, a pioneer in microclimatic design issues during the 1930s–1940s), the city's inclination with regard to the north depended mainly on both the main winds' directions and sun exposure.

His studies on Roman *castra* (Roman military camps) advanced interesting and important theories on Ancient knowledge regarding a city's healthiness. In particular, Vinaccia's research highlighted that the north–south-oriented plan was extremely rare, «perhaps due more to fortuity than to sacred precepts' observance». Conversely, a plan's disposition on territory depended mainly on the local necessity to avoid 'annoying' and insalubrious winds inside the urban areas. The author's theories found their foundation in ancient manuscripts, particularly in Vitruvius, who stated that: «[cities] will be well built if winds are wisely kept out from squares and streets; [winds] which offended if cold, vitiate if hot, harm if wet» [40]. The Ancients knowledge of winds was proved by the construction of a bronze wind rose, representing a triton, on which were depicted the 12 main winds. The station was probably built by Flavius Eutropius, during Valentinian II's empire. According to the winds' mythological representation, the 'annoying' winds came from north/north-westerly directions (cold winds), whereas the 'unhealthy' ones arrived from the south-south-east (usually, hot wet winds) (**Figure 2**).

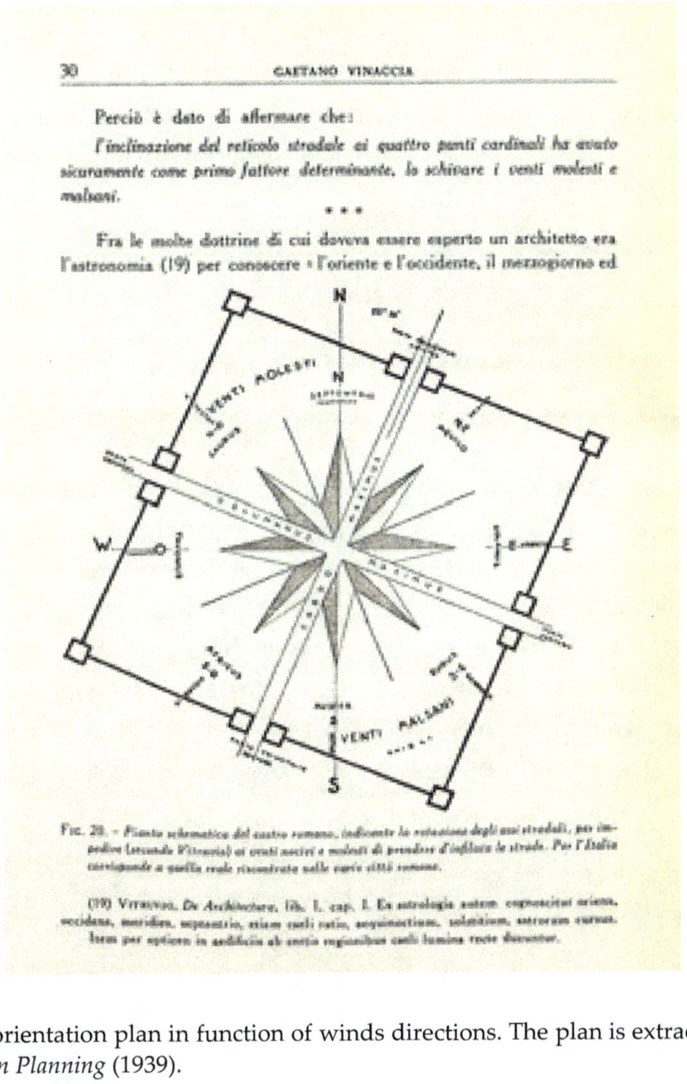

30 GAETANO VINACCIA

Perciò è dato di affermare che:

l'inclinazione del reticolo stradale ai quattro punti cardinali ha avuto sicuramente come primo fattore determinante, lo schivare i venti molesti e malsani.

* * *

Fra le molte dottrine di cui doveva essere esperto un architetto era l'astronomia (19) per conoscere « l'oriente e l'occidente, il mezzogiorno ed

Fig. 28. - [illegible]

(19) Vitruvio, *De Architectura*, lib. I, cap. I. [illegible]

Figure 2. Roman castrum orientation plan in function of winds directions. The plan is extracted from Vinaccia's *Orientation issue in Ancient Roman Planning* (1939).

Plans' focus on wind conditions needed to be coupled with local solar dynamics. In regard to Roman castra located in Italy, Vinaccia affirmed that their orientation, useful to prevent the built-up area from injurious winds, also aimed at assuring the best exposure to the sun during the entire year. The rotation of almost 22°N, which, according to Vinaccia, corresponds with the average orientation measured for Italian cases in the study, allowed sunrays to reach all the buildings' facades, including the northern ones. This 'equisolare' orientation, which will be better investigated in the author's successive works, permitted designers to rationalize solar radiation prorating it in function of seasonal needs. This theory finds its roots in ancient essays, in which Latin authors (Varrone, Columella, Cato and Vitruvius) related the buildings' dispositions with local «orienting urban points», which corresponded to the horizon's sunrise and sunset points in solstices and equinoxes. The depth of past studies shows the Ancients' advanced knowledge on both the sun's path (during seasons and at different latitudes) and solar radiation intensity, as a function of solar height and the hours of the day.

Currently, the Department of Architecture at the University of Cagliari (Prof. G. M. Chiri) is studying the effectiveness of Vinaccia's theories on Roman castra's orientation. At the moment, the research has examined the fluid-dynamic behaviour of several Italian case studies, analysing the relationship between urban forms and local wind conditions. Analyses have taken into account the urban grids' orientations and the main wind directions during the winter and summer seasons. CFD calculations have been developed in collaboration with the CONSELF Company (https://conself.com/). A short example, representative of the work, is summarized in **Box n.1**.

Box n.1

***Luca* - Roman Castrum**

(authors: G.M. Chiri, I. Giovagnorio, and D. Usai)

According to Livy, the Roman castrum of *Luca* (Lucca, Italy) was founded in 180 B.C. The name root, '*luk*' ('marshy place'), which reveals its Celtic-Ligur origin, originated from the environmental context in which it arose (1). Urban form and location were influenced by the Serchio River, which still flows today to the northern part of the modern city. Thanks to its position, *Luca* was a relevant road hub until the Middle Ages. The Roman city was quadrilateral —it shows just a little irregularity on the northern side—and its east–west fronts measured around 500 m, whereas the north–south ones were around 650 m. The urban grid, constituted by secondary orthogonal streets, defined rectangular blocks (*insulae*) of 105 × 120 m.

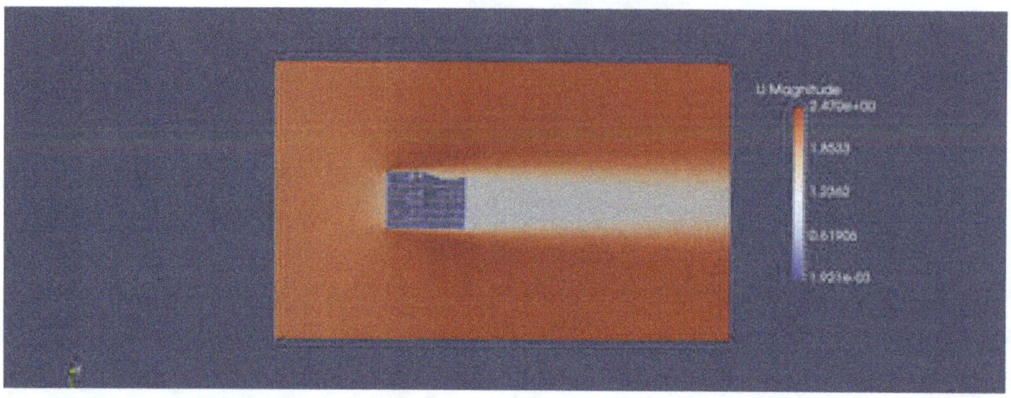

Figure 3. *Luca*'s wind flows in summertime.

According to results gathered in 2005 by Frisia et al. (2) on paleoclimatology, and quoted by Chen et al. (3) in 2011, «the "Roman Classical Period" temperatures were similar or slightly higher than those of today with the highest temperatures reached between around 400 BC and 0 AD». This correspondence supposes similar data to those of contemporary conditions. In this light, the validation of Vinaccia's statements has been done, taking into consideration Lucca's current weather data (Meteonorm database source)—in particular, the winds' main directions and monthly values in the wintertime (≈90 N and ≈3 m/s) and the summertime (≈270 N and ≈2.5 m/s).

Analyses on wind flows have been developed with the support of the CONSELF company (CFD software); the results confirm Vinaccia's theory. In fact, Lucca's environmental condition as a 'marshy place' led the Romans to orient the urban grid in the direction of the main winds, in order to remove insalubrious humidity throughout the year (**Figure 3**). Contrary to common belief, Lucca's north–south orientation is not due to religious or astronomical precepts but to practical, healthy needs.

3.3. The Renaissance treatises

Knowledge of Latin was rediscovered during the Renaissance, following the humanistic culture's diffusion. Architecture acquired an intellectual rigour typical of the liberal arts, such as literature and science. Its progress, coupled with advancements in drawing techniques, contributed to transforming design practices in a scientific, exact procedure [41]. The birth of perspective allowed designers to describe objects' proportional features directly (their shapes and mutual positions) and, indirectly, their dimensions and physical characteristics. Relationships among urban and architectural elements took inspiration mainly from Roman models; nevertheless, according to Benevolo, «thus, Renaissance architecture achieves its model of proportionality and regularity in single buildings, whereas it is not able to establish, or change, a whole city». However, L.B. Alberti's urban theories, contained in *De Re Aedificatoria,* pose the basis for the planning experiences of the fifteenth century. Since at its origins were Latin's principles, in particular those of Vitruvius, also the Renaissance theories on design practices pointed attention to local microclimatic characteristics [42]. According to Benevolo [41], European colonies in American territories are the most significant urban planning examples of the sixteenth century. New cities employed the urban grid, usually defined by square blocks. This model was codified by Philip II into the first town-planning act of the modern era, in 1573. The law included recommendations to protect the public square from the main winds. Microclimatic advantages of the old grids, compared to the successive Jeffersonian one, have been studied by R. Knowles [28] for the city of Los Angeles. The author highlighted the relationship between Spanish grids' orientation (nearly 45°) and sea breezes, and the grid best performances in relation to the sun's access throughout the year.

In Renaissance examples, bioclimatic buildings were more common. They exploited passive cooling and heating systems, using natural ventilation and solar radiation. Sirocco rooms in Sicily [43, 44], Rafaello's Villa Madama (1517) and Vatican Loggias [45], in Rome, and the Costozza's Villas (Vicenza) [46], appear to be particularly significant examples. The latter was quoted by Palladio in his essay *I quattro libri dell'architettura* (*The Four Books on Architectures*), one of the most important Renaissance treatises, which was published in 1570 [47]. The ancient relationship between architecture and local environmental conditions also permeates Palladio's work. Several important references are reported in the treatise concerning both the building and the urban issues. Particularly interesting are Palladio's considerations on the villas' sites (Chapter XI, Book II) and on urban streets (Chapter II, Book III), where reflections extend from buildings to the surrounding areas. Concerning the latter, the author suggests orienting and dimensioning them as functions of the sun's access and local wind conditions, in strict coherence with Vitruvius's recommendations.

3.4. Bioclimatism in the European Modern Movement

In the early nineteenth century, cities' awful conditions after the Industrial Revolution led to the Hygienism movement. Hygienists condemned the squalid, crowded working-class slums, considering the absence of fresh air and sunlight the main reason for lethal diseases' development [48–50]. This belief mobilized several studies, besides concerning disciplines correlated to the architectural issues (such as medicine, physics, sociology), which proved the absolute efficacy of sunlight and solar radiation on living spaces' healthiness. Starting from the end of

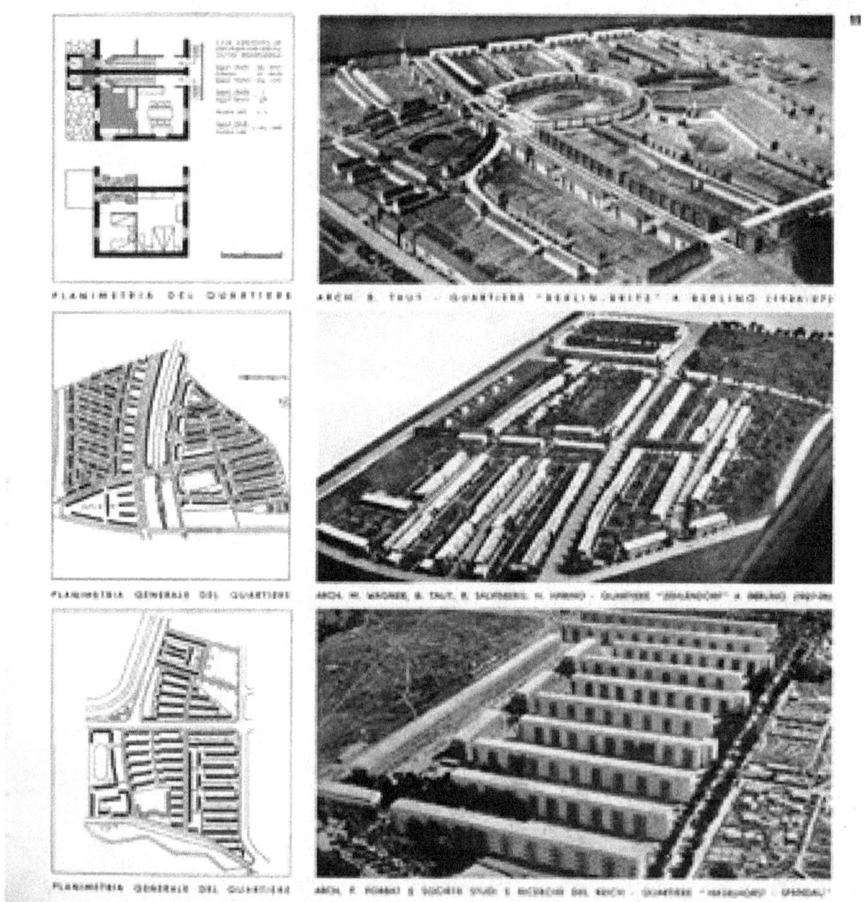

Figure 4. Examples of German modern 'row-houses'. Extraxt from Diotallevi, Marescotti, *Aspetti e Problemi della Casa Popolare.* Casabella n.164 (1941), p. 11.

the nineteenth century «architects and planners began to study the question of solar orientation more scientifically» [51], with the exception of a few rare previous examples, such as Port Sunlight in England, built in 1860 [51, 52]. A broad classification of European solar studies on buildings and the grid street orientation had been developed in Harzallah [50] and Montavon's [49] doctoral theses. The growing interest in buildings' salubrity fed the international architectural debate, focusing initially on building typological research (CIAM II, Frankfurt am Main, 1929) [53, 52, 56, 58] and, later, widening the scope to the city (CIAM III, Bruxelles, 1930) [54]. During the third CIAM congress, Siegfried Gideon [55] defined the goal of the so-called Modern Movement as that of reaching the most efficient formulation for a typological scheme for building on a different scale. The proposal for new neighbourhoods based on the *Zeilenbau Plan*—linear high-rise blocks— which was developed in the early 1920s in Germany, became the standard (**Figure 4**).

The influence of the German avant-garde on urban solar studies is demonstrated by its early research and urban experiences. In 1824, Faust expounded in his *Sonnenbaulehre* (building orientation theory) [56] the plan for a solar city. According to Plessner [57]), Vitruvius largely inspired Faust. His Sonnenbau system aimed at providing as much sun as possible to houses by planning settlements on a north–south-oriented grid and by ensuring the correct distance between the blocks. The Sonnenbau theory was supported by the Bayern architect, Gustav Vorherr (1778–1847), who designed in 1818–1821 the so-called "Sun Road" [58]. The road, placed on the border between the ancient city centres of Munich and Ludwig, is strictly south oriented. Interesting theories on buildings' solar orientation were also published in 1879 by Franz Knauff [59] and Alfred Vogt [60]. In 1919, Theodor Fisher (1862–1938) realized the Alte Heide plan, in which each block is at a distance from other ones by twice its height, in order to prevent façade shading. Five years later, Otto Heasler (1880–1962) developed the principle in the Georgsgarten Siedlung. This rule, later called the Heiligenthal rule after R.F. Heiligenthal's studies [61] on German city planning, became the *Neue Sachlichkeit*'s standard approach for housing-climate control between 1925 and 1933, after attracting the enthusiasm of the international architectural community. The linear high-rise blocks were south–north oriented in order to ensure maximum and equal insulation for the apartments. Gropius's diagrams aimed to define a new universal typology for new settlements and were applied in «numerous long, narrow apartment buildings that soon dotted the German landscape» [51]. The history of architecture reported several examples: W. Gropius's Daamerstock Siedlung (1929), the Siedlungen of E. May in Frankfurt (1925–1930) and the Siemensstadt Colony of H. Scharoun in Berlín (1929–1931). In 1927, Ludwig Hilberseimer published his *Großstadtarchitektur* [52], starting from a criticism of the Siedlung model. In his plan for a 'vertical city', the urban grid is oriented as a function of sun exposure and the dimensions of both streets and blocks adhered to ventilation and sunlight requirements in order to ensure the provision of space, air and sun to each house. In his studies, Hilberseimer conferred great importance to the block's arrangement, defining both different typologies in terms of orientation [34], building density and shape in relation to the sun's access at different latitudes [49].

France also developed pioneering studies strictly connected to solar urban design. According to Butti and Perlin, «The possibility of extending the same benefits [spacious, sunlit working-class communities] to urban residents inspired Augustin Rey, a French housing official, to investigate the feasibility of comprehensive urban solar planning» [51]. In 1912, his studies focused on the minimum blocks' distance in order to avoid winter shadows among buildings. Rey analysed the winter sun's path for several cities and correlated the different land areas' consumptions in relation to north–south and east–west blocks' orientations. His knowledge of the daily difference in the time it took to reach the maximum solar radiation and the maximum air temperature—*la vague thermique*— led Rey together with Barde and Pidoux, to develop the heliothermic theory [62], *«qui prétend rendre compte de «l'échauffement» des bâtiments en tenant compte à la fois de leur exposition au soleil et de la température de l'air»* [63]. The wide dissemination of the heliothermic axis influenced the work of several important designers, such as Le Corbusier, who presented at the third CIAM in Brussels his plan for the Ville Radieuse, which was oriented by 19 northeast degrees [64]. The axis's effectiveness was disputed during the 1940s; it opposed the "hygienists", the partisans of east–west exposure, to the "climatologists", who were advocates of southern exposure, among whom was Vinaccia (1943), who counterposed his *equisolare* orientation theory [65]. Even though Vinaccia did not have available contemporary tools to demonstrate the correctness of his opinion, it is interesting to highlight (as recent research has confirmed) the inefficacy of Rey's theory [63, 66].

The *Cité Industrielle* (1917) of Tony Garnier [67] was also designed by the author as a function of healthy principles. According to Butti [51], he was «heavily influenced by studies of ancient Greek city planning, the young designer [...] bore a close resemblance to Olynthus and Priene». Long rectangular east–west oriented blocks recall the ancient Greek organization, as well as buildings' arrangement, whose form «was to achieve good ventilation and high levels of sunlight into all homes».

Although Italy shows representative examples, such as Vinaccia's works, the international debate on urban healthiness arrived several years after the previous experiences. Irenio Diotallevi and Franco Marescotti are still often considered the Italian forerunners for the topic of building-related illnesses. In *Costruzioni-Casabella,* they published some evidence for the relation between wellness and housing. These articles, which were grouped together in *Ordine e destino della casa popolare* (1941) [68], aim to directly link illness and the shape of buildings, focusing on the lack of insulation and ventilation as the main factors for the onset of disease in the working class.

Within the national architectural debate, the Roman architect Gaetano Vinaccia attempted to use several publications to discuss the relationship between urban form and local microclimates. Because of his status as a minor architect, his biography has not been studied in depth and only modest information is available [69]. In examining his productions, we can recognize four different phases [70]. In the first period, from 1919 until 1926–1927, he applies his knowledge to the field of architecture, achieving only mediocre success. The second, from 1926 to 1930, which corresponds to his studies in Freiburg where he graduated in civil engineering, can be considered as the more fruitful experience of his life, due to the opportunity to be in contact with the most advanced research in architecture, meteorology and civil engineering. The third occurs during his stay in Rome, when, as a board member of the architectural magazine *Case d'Oggi,* he tried to influence the debate on typology, which was led by rationalist architects worldwide. The last phase was during his attempt to lead the field towards an Italian approach to urban microclimate design and closed with the publication of his most frequently cited work, *La Città di Domani, Come il clima plasma la forma urbana e l'architettura: la sanità e l'igiene cittadina, Vol. 1* (1943) [65]. This work, which is the first complete treatise on the matter, marks him out as an absolute pioneer, even if his influence on architecture and urbanism has not been considerable, due to the unfavourable context in which his studies evolved.

Although most of Vinaccia's theories were already known in distinct international scientific sectors (climatology, physics, astronomy, etc.) he was perhaps the first to organize these into a systematic approach, also thanks to his education as an architect, which contributed to a more humanistic idea of architecture and planning. In this light, it is important to highlight the author's attempt to found a new scientific discipline, called *Polisclimatology*, in-between planning and microclimatology. His nature as a 'polyhedral' architect, able to manage design's various aspects that were related to distinct disciplines, made him an absolute innovator and his ideas extremely contemporary in our own time.

4. Contemporary experimentations

After World War II, the illusory belief in the technological system's supremacy led architects to forget their past environmental tradition. This trend continued until the 1970s–1980s, when the international oil crisis and environmental disasters (global warming, 1986; the hole in the ozone layer, 1985; the Chernobyl disaster, 1986; etc.) showed clearly the environment's weaknesses. This induced technicians to rethink the global development model in favour of a more sustainable one. Bioclimatism in architectural and urban design practices, which had in Rudofsky's vernacular research (1964) [71] and Olgyay's architectural regionalism (1963) [72] two anticipatory events, caught on again. Scientific and technological progress that had occurred during the last few decades, allowed integration of the design process with important external 'environmental' contributions of interrelated disciplines (technical physics, fluid dynamics, microclimatology, computer engineering, etc.). According to the Bjarke Ingels Group (BIG), «[computer information models] allow us to shift the ultimate performance of a building away from the mechanical room and back into the permanent attributes of the design» [73]. Complexities resulting from this interdisciplinary perspective call for an update to the current process without being overcome by them.

On this topic, Monserrato's master plan tests a methodology (**Figure 5**) through which to integrate both environmental data and analyses, starting from the design's initial phases. It has been supported by Heliodon and ENVI-met software that act as useful 'feedback' tools able to verify qualitatively the environmental behaviour of the project's concept. In particular, the environmental analyses focused on the microclimatic consequences of urban form, taking into consideration different climatic data (air temperature, relative humidity, wind directions and velocities, hours of sunlight and direct solar radiation) and form parameters (aspect ratio and the sky view factor). Analyses were conducted in relation to winter and summer solstices, characterized by the most extreme environmental conditions during the year. The initial urban proposal has been gradually modified and re-evaluated several times in relation to the main criticalities that have emerged from analyses' results and bio-climatic diagrams' information (Olgyay). Nevertheless, modifications respect the master plan fundamentals and collaborate in the improvement of its overall performance.

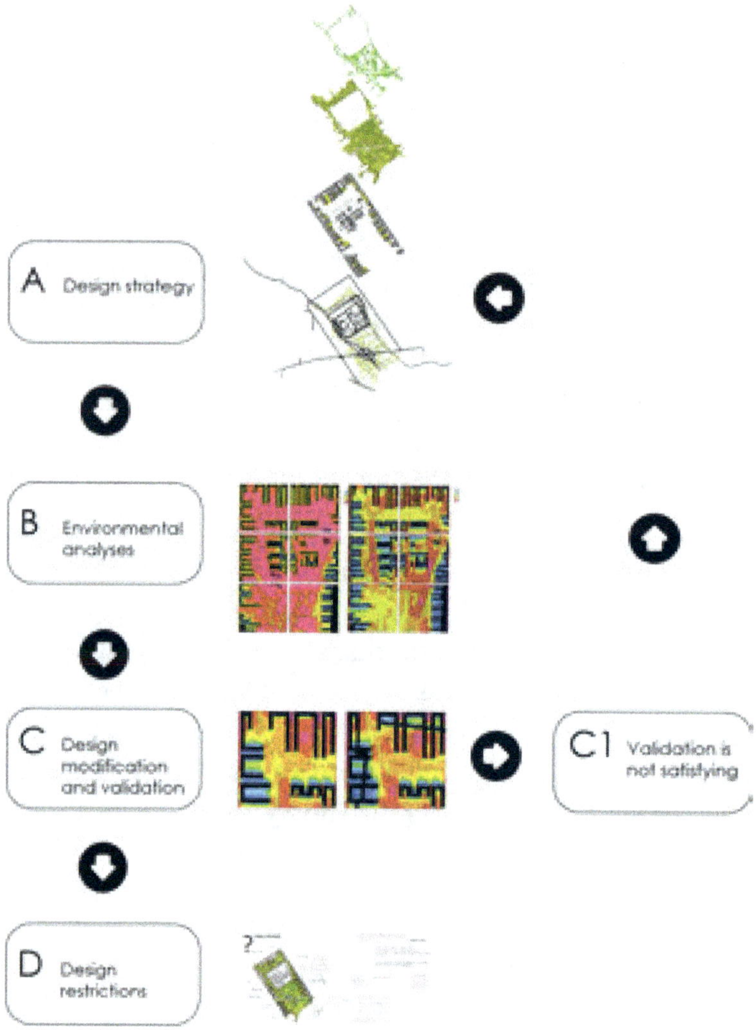

Figure 5. Diagram of design methodology.

The case study concerns a new urban expansion around the Academic Citadel area, in Monserrato (Sardinia, Italy). The main goals of the master plan were the reconnection of the University with the city centre, starkly separated by Highway 554; the urban sprawl's containment and, finally, the respect of local agricultural tradition. With these aims, the project designed an anti-sprawl territorial system, which encircles the existing Citadel with a new urban district (1.4 km^2) linking it to the north-west city's edge (**Figure 6**). The blocks are arranged on the perimeter of the area along the existing position that is already in common with the historical city centre, the Citadel and the waterways system. The Citadel and the new expansion are interconnected by a wide central park, towards which the 'c-shaped' blocks open their inner courts. Their connection is achieved through a podium, which bends back on itself, defining two different public spaces. The park's layout makes a direct reference to the previous rural pattern, with past tracks of land's plots, which host pedestrian and cycle paths bordering different green areas (lawns, gardens, etc.) (**Figure 7**). The blocks' dimensions are contained between 50 and 200 m, and the maximum height remains constant, opposing the increasing altitude of the terrain. Functions and the infrastructure network are designed in order to assure mixed use within the 400 m.

Figure 6. View of the master plan.

Environmental analyses evaluated environmental criticalities, taking into account man's thermal comfort in open spaces. Results recognized that the worst performances occurred during the summer season, when north-west blocks reached high wind velocities and high solar radiation values in courtyards, which were exposed to direct sunrays for most of the day. Uncomfortable conditions were also confirmed by Olgyay's diagrams, in which this cluster of blocks moves away from average seasonal values.

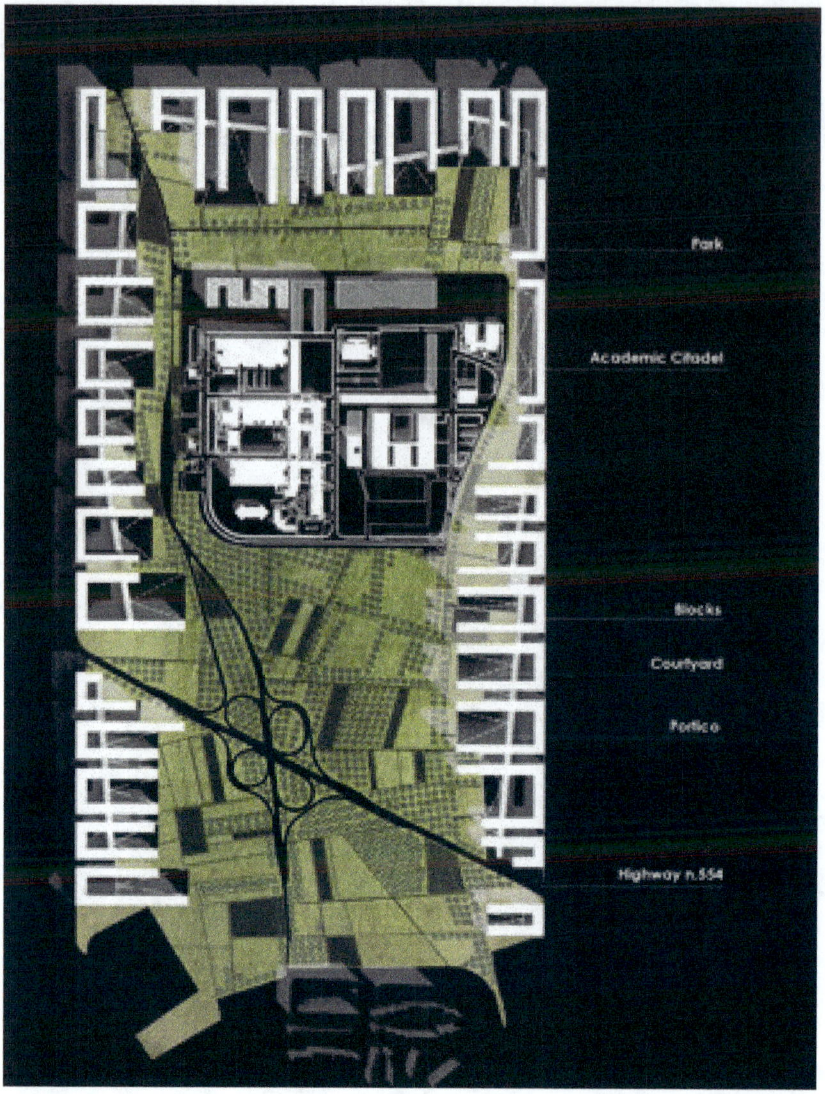

Figure 7. Monserrato plan.

In order to mitigate environmental weaknesses, the initial design has been modified, focusing on the northern part. Specifically, modifications concerned the blocks' height, which was raised in order to protect courtyards from the cold mistral, the introduction of a portico in each courtyard aimed at increasing shadows, and, finally, greenery's variation and implementation. The 'updated' master plan has been tested again in order to verify the variations' efficacies, confirming the performances' improvement. The additional vegetation, together with the portico, collaborate in reducing wind velocities without obstructing airflows. At the same time, the insertion of the latter increased shadows on courtyards' surfaces, reducing direct radiation by 20% and sun hours by 25–30% (**Figure 8**—above).

Referring to the transformation process adopted in several contemporary urban designs, such as the ZAC Bercy (Paris) by the architect J.P. Buffi [74], the work defined a set of restrictions in order to preserve the master plan's overall strategies and shape. Typological and morphological rules did not represent such blocks' final configurations, but they did have the task of explaining clearly to designers the spatial relationships among parts, guiding their successive work on single blocks (**Figure 8**—below). The restrictions list of Monserrato metadesign coupled general rules with technical datasheets that referred to specific blocks. The former are structured in six main sections (1. edges; 2. blocks; 3. special blocks; 4. pedestrian routes and courtyards; 5. park and gardens; 6. roads), whereas the latter include special, additional rules combined with explanatory drawings.

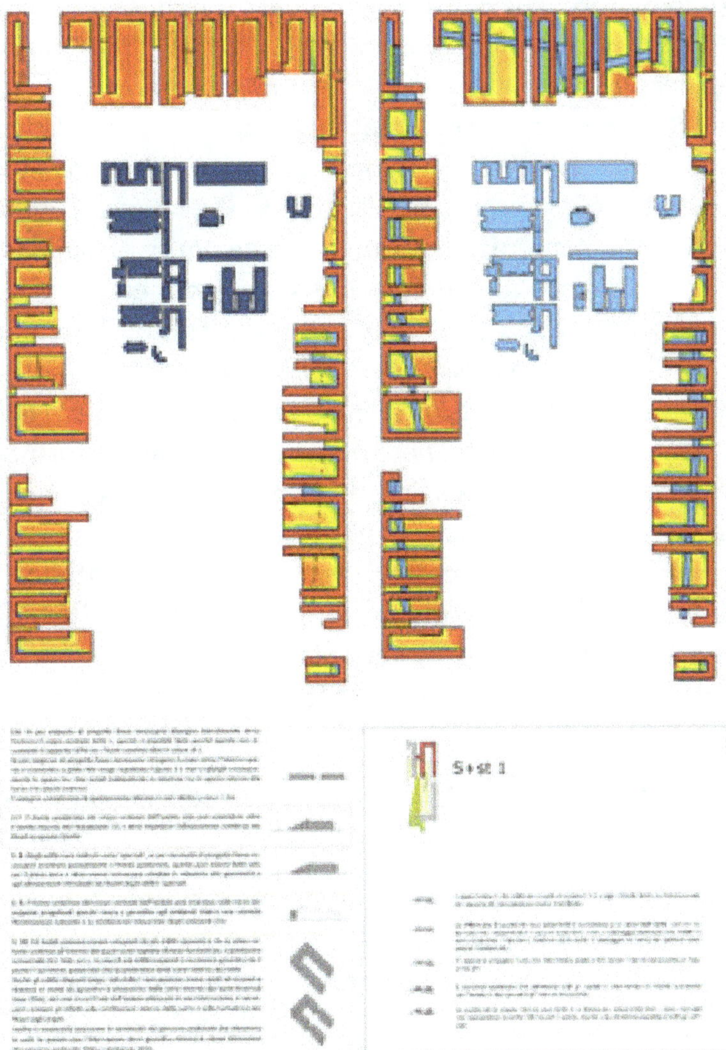

Figure 8. Design's comparison of sun hours analysis (above) and examples of design general rules (left) and datasheet (right) (below).

5. Conclusion

According to most current scholars, goals in sustainable architecture need to be approached via the urban dimension. The necessity of this 'change of scale' has been proved, in the last few years, in several national and international scientific works, in which authors highlighted the benefits that are achievable—in relation to both climate change goals and energy issues.

In order to overcome sustainable challenges, urban design, which falls in between city planning and architectural design, becomes the appropriate tool through which operate, thanks to its ability to solve 'upstream' the buildings' environmental and energy deficiencies. The strategy was common in ancient times, when bioclimatic urban solutions were applied, both to mitigate extreme environmental conditions and to exploit passively the sun's heat in wintertime. Examples of 'solar cities' across the history of architecture until World War II, after which technology-system diffusion and the functions' specializations, were deemed less interesting in environmental practices. Despite the fact that bioclimatism partially caught on during the energy and environmental crises of the 1970s and 1980s, it was mainly applied to single buildings, neglecting urban theories. Today, when important research has shown the limits of strategies related to energy consumption reduction in single buildings and favours broader policies, ancient lessons have acquired relevance. In particular, the 'polyhedral' figures of past architects have gained in prominence, as they have provided us with extensive interdisciplinary knowledge. A significant example is the figure of Gaetano Vinaccia, an Italian architect and pioneer in urban microclimate design. Although he is classified as a minor architect and only a little information is available regarding his biography, his most cited work, *La Città di Domani, Come il clima plasma la forma urbana e l'architettura: la sanità e l'igiene cittadina, Vol. 1* (1943), can be considered the first complete treatise on urban environmental design and marks him out as an absolute innovator. These urban environmental principles, which allowed past technicians to manage urban shapes and functions without renouncing quality of space, have to be recovered by contemporary architects. Scientific progress over the last few decades allows architects and planners to reintegrate the project process with the important contributions of external disciplines. Complexities resulting from this multidisciplinary perspective call for an update to the current process, without it being overcome by those complexities. In this light, the methodology tested through the Monserrato case study offers a point of view on current design practices.

Acknowledgements

We would like to express our gratitude to Cesare Silvi, who shared his information on Gaetano Vinaccia's biography, and to the CONSELF Company for helping us in CFD calculations and analyses.

Author details

Ilaria Giovagnorio* and Giovanni M. Chiri

*Address all correspondence to: ilaria.giovagnorio@gmail.com

DICAAR, University of Cagliari, Cagliari, Sardinia, Italy

References

[1] European Commission. The Vancouver Declaration On Human Settlements. From the Report of Habitat - United Nation Conference on Human Settlements [Internet]. Vancouver, Canada; 1976. Available from: http://unhabitat.org/wp-content/uploads/2014/07/The_Vancouver_Declaration_1976.pdf

[2] Larsen L., Rajkovich N., Leighton C., McCoy K., Calhoun K, Mallen E, et al. Green Building and Climate Resilience. Understanding impacts and preparing for changing conditions. University of Michigan; U.S. Green Building Council; 2011. 260 p.

[3] Hachema C, Athienitisb A, Fazio P. Evaluation of energy supply and demand in solar neighborhood. Energy Build. 2012;Volume 49:335-47.

[4] Oke TR. Boundary Layer Climates. Methuen; 1987. 435 p.

[5] Landsberg H. The Urban Climate. New York, USA: Academic Press; 1981.

[6] Givoni B. Urban design in different climates. World Meteorol Organ WMO/TD. 1989;(346).

[7] Ratti C, Baker N, Steemers K. Energy consumption and urban texture. Energy Build. 2005;37(7):762-76.

[8] Steemers K, Ratti C, Raydan D. Building form and environmental performance: Archetypes, analysis and an arid climate. Energy Build. 2003;35(1):49-59.

[9] Cheng V, Steemers K, Montavon M, Compagnon R. Urban Form, Density and Solar Potential. CLEVER DESIGN AND AFFORDABLE COMFORT A Challenge for Low Energy Architecture and Urban Planning PLEA2006 - The 23rd Conference on Passive and Low Energy Architecture. Geneva, SWITZERLAND; 2006.

[10] Nikolopoulou M, Baker N, Steemers K. THERMAL COMFORT IN OUTDOOR URBAN SPACES: UNDERSTANDING THE HUMAN PARAMETER. Sol Energy. 2001;70(3): 227-35

[11] Gupta V. Solar radiation and urban design for hot climates. Environ Plan B Plan Des. 1984;11:435-54.

[12] Gupta V. Thermal efficiency of building clusters: an index for nonair-conditioned buildings in hot climates. In: Hawkes D, editor. Energy and Urban Built Form. Oxford, UK: Butterworths, UK; 1987.

[13] Compagnon R. Solar and daylight availability in the urban fabric. Energy Build. 2004;36(4):321-8.

[14] Knowles RL. The solar envelope: Its meaning for energy and buildings. Energy Build. 2003;35(1):15-25.

[15] Ng E. Designin for Urban Ventilation. In: Ng E, editor. Designing High Density Cities - For Social and Environmental Sustainability. London, UK: Earthscan Publications Ltd.; 2009. p. 119-36.

[16] Ng E, Yuan C, Fung JC, Ren C, Chen L. Improving the wind environment in high-density cities by understanding urban morphology and surface roughness: A study in Hong Kong. Landsc Urban Plan. 2011;101(1):59-74.

[17] Allard F, Ghiaus C. Natural Ventilation in the Urban Environment: Assessment and Design. London, UK: Routledge; 2005. 266 p.

[18] Allard F, Ghiaus C, Szucs A. Naturai Ventilation in High-Density Cities. In: Ng E, editor. Designing High Density Cities - For Social and Environmental Sustainability. London, UK: Earthscan Publications Ltd.; 2009. p. 137-62.

[19] Tablada A, De Troyer F, Blocken B, Carmeliet J, Verschure H. On natural ventilation and thermal comfort in compact urban environments - the Old Havana case. Build Environ. 2009;44(9):1943-58.

[20] Moonen P, Defraeye T, Dorer V, Blocken B, Carmeliet J. Urban Physics: Effect of the micro-climate on comfort, health and energy demand. Front Archit Res. 2012;1(3): 197-228.

[21] Panagiotou I, Neophytou MK-A, Hamlyn D, Britter RE. City breathability as quantified by the exchange velocity and its spatial variation in real inhomogeneous urban geometries: An example from central London urban area. Sci Total Environ. 2013;442:466-77.

[22] Towards an ICT Infrastructure for Energy-Efficient Buildings and Neighbourhoods for Carbon-Neutral Cities. Final Report on the Advisory Group Workshop [Internet]. Brussels; 2010. Available from: http://ec.europa.eu/information_society/activities/sustainable_growth/cities/index_en.htm

[23] Knowledge NEU, CIPU. Certification Systems for Sustainable Urban Neighbourhoods [Internet]. 2014. Available from: http://www.eukn.eu/events/policy-labs/certification-systems-for-sustainable-urban-neighbourhoods

[24] Carlos GD, Correia M, Viana D, Merten J. Settlements morphology. In: Correia M, Dipasquale L, Mecca S, editors. VERSUS: HERITAGE FOR TOMORROW Vernacular Knowledge for Sustainable Architecture [Internet]. Florence, Italy: Firenze University Press; 2014. p. 90-9. Available from: http://www.esg.pt/versus/versus_heritage_for_tomorrow.pdf

[25] De Pascali P. Città ed energia. La valenza energetica dell'organizzazione insediativa. Franco Angeli Editore; 2008. 368 p.

[26] Picone A. Lo sviluppo sostenibile dell'Oasi di Siwa, un processo corale. Costruire sostenibile Il Mediterraneo. Alinea Editrice; 2001. p. 94-9.

[27] Picone A. L'architettura dell'Oasi di Siwa, tra natura, clima e tradizione. Area. 2001;58:86-95.

[28] Knowles RL. Sun Rhythm Form. Cambridge, Mass: The MIT Press; 1985. 304 p.

[29] Knowles RL. Energy and Form: Ecological Approach to Urban Growth. MIT Press; 1974. 198 p.

[30] Butti K, Perlin J. Solar Heating in Early America. A Golden Thread: 2500 Years of Solar Architecture and Technology. New York, USA: Van Nostrand Reinhold; 1980. p. 172-9.

[31] Vegas F, Mileto C, Cristini V, Checa JRR. Underground cities. In: Correia M, Dipasquale L, Mecca S, editors. VERSUS: HERITAGE FOR TOMORROW Vernacular Knowledge for Sustainable Architecture. Florence, Italy: Firenze University Press; 2014. p. 114-27.

[32] Caltabiano I. Consapevolezza energetica nelle costruzioni tradizionali in area mediterranea. Ingegno e natura al servizio dell'abitare. In: Biondi B, editor. 1st International Research Seminar on Architectural Heritage and sustainable development on small and medium cities in south mediterranean regions Results and strategies of research and cooperation. Pisa, Italy: Edizioni ETS; 2005. p. 470-8.

[33] Secchi B. La Città Normale. In: Mittner D, editor. La città reticolare e il progetto moderno. CittàStudiEdizioni; 2008. p. 47-58.

[34] Los S. CITTA' SOLARI dal passato al futuro. IUAV. (42):1-16.

[35] Benevolo L. Storia della Città. Vol 1.La Città Antica. 2nd editio. Roma, Italy: Laterza Editori; 2006. 319 p.

[36] Castagnoli F. Ippodamo di Mileto e l'Urbanistica a Pianta Ortogonale. Roma: De Luca Editore; 1956. 107 p.

[37] Von Gerkan A. Griechische Stadteanlagen. Berlin Leipzig; 1924.

[38] Butti K, Perlin J. Solar Architecture in Ancient Greece. A Golden Thread: 2500 Years of Solar Architecture and Technology. New York, USA: Van Nostrand Reinhold; 1980. p. 289.

[39] Vinaccia G. Il Problema dell'Orientamento nell'Urbanistica dell'Antica Roma. Roma: Istituto di Studi Romani; 1939. 42 p.

[40] Vitruvio. De Architettura. Torino, Italy: Einaudi; 1997.

[41] Benevolo L. Storia della Città. Vol.3. La Città Moderna. Roma, Italy: Laterza Editori; 2006.

[42] Benevolo L. Storia dell'architettura del Rinascimento. Roma, Italy: Laterza Editori; 2002.

[43] Dipasquale L. Sirocco Room. In: Correia M, Dipasquale L, Mecca S, editors. VERSUS: HERITAGE FOR TOMORROW Vernacular Knowledge for Sustainable Architecture. Firenze, Italy: Firenze University Press; 2014. p. 207.

[44] Saeli M, Saeli E. Analytical studies of the Sirocco room of Villa Naselli-Ambleri: A XVI century passive cooling structure in Palermo (Sicily). J Cult Herit [Internet]. 2015;16(3): 344-51. Available from: http://www.sciencedirect.com/science/article/pii/S1296207414000867

[45] Jankovich B. CLIMA E PROGETTO Note sulla progettazione bioclimatica degli spazi architettonici interni ed esterni. Firenze, Italy: Edizioni Medicea; 1990.

[46] Dell'Osso R. L'architettura della villa. Maggioli Editore; 2008. 185 p.

[47] Palladio A. I quattro libri dell'Architettura. Milano, Italy: Hoepli; 1990.

[48] Benevolo L. Storia della città. Vol.4. La Città Contemporanea. Roma, Italy: Laterza Editori; 2006. 332 p.

[49] Montavon M. Optimisation of Urban Form by Evaluation of the Solar Potential. Ecole Polytechnique Federale, Lausanne, France; 2010.

[50] Harzallah A. Émergence et évolution des préconisations solaires dans les théories architecturales et urbaines en France, de la seconde moitié du XIXe siècle à la deuxième guerre mondiale. University of Nantes; 2007.

[51] Butti K, Perlin J. Solar Communities of Europe. A Golden Thread: 2500 Years of Solar Architecture and Technology. New York, USA: Van Nostrand Reinhold; 1980. p. 289.

[52] Hilberseimer L. Großstadtarchitektur. L'Archtiettura della Grande Città. Napoli, Italy: Clean; 1998.

[53] The International Congress of Modern Architecture (CIAM) focused on the work of Hannes Meyer, Mart Stam, Walter Gropius, Erns May, and Alexander Klein on Existenzminimum ("minimum dwelling"), aimed at reducing building costs (including the cost of land co.

[54] Aymonino C. L'Abitazione Razionale, Atti dei Congressi CIAM 1929-1930. 4th ed. Venezia, Italy: Marsilio; 1970.

[55] Gideon S. Spazio, Tempo Architettura. 2nd ed. Milano, Italy: Hoepli; 1987.

[56] Faust BC. Andeutungen über das Bauen der Häuser und Städte zur Sonne. Hannover, Germany: Hahn'sche Hofbuchhandlung; 1829.

[57] Plessner H. Die Sonnenbaulehre des Dr. Bernhardt Christoph Faust: Ein Beitrag zur Geschichte der Hygiene des Staàdtebaus. Technische Hochschule, Berlin, Germany; 1933.

[58] Kratzer FA. Das Stadtklima. 2nd ed. Braunschweig: Vieweg; 1956.

[59] Knauff F. Da neue akademische Krankenhaus in Heidelberg. München, Germany; 1879.

[60] Vogt A. Über die Richtung der städtischen Straßen nach der Himmelsgegend und das Verhältnis ihrer Breite zur Häuserhöhe. Z Biol. 1879;

[61] Heiligenthal RF. Deutsche Städtebau. 1st ed. Heidelberg, Germany: Carl Winter; 1921.

[62] Rey AA, Pidoux J, Barde C. La Science des Plans de Villes. Paris, France: © Dunod; 1928.

[63] Harzallah A, Siret D, Monin E, Bouyer J. Controverses autour de l'axe héliothermique: L'apport de la simulation physique à l'analyse des théories urbaines. INHA [Internet]. 2014; Available from: http://inha.revues.org/2509

[64] LeCorbusier. La Ville Radieuse: element d'une doctrine d'urbanisme pour l'equipment de la Civilization Machinist. De L'Architecture D'Aujourd'Hui. 1935;

[65] Vinaccia G. La Città di Domani. Vol1. Come il Clima Plasma la Forma Urbana e l'Architettura. Roma: Fratelli Palombi Editore; 1943. 155 p.

[66] Cheng V, Steemers K, Montavon M, Compagnon R. "La Ville Radieuse" by Le Corbusier once again a case study. CLEVER DESIGN AND AFFORDABLE COMFORT A Challenge for Low Energy Architecture and Urban Planning PLEA2006 - The 23rd Conference on Passive and Low Energy Architecture. 2006.

[67] Garnier T. Una Città Industriale. Milano, Italy: Jaca Book; 1990.

[68] Diotallevi I, Marescotti F. Ordine e destino della casa popolare. Milano, Italy: Editoriale Domus; 1941. 103 p.

[69] Silvi C. Solar Building Practices and Urban Planning in the Work of Gaetano Vinaccia (1889-1971). Proceedings of the 2nd International Solar Cities Congress. Oxford, UK; 2006.

[70] Chiri GM, Giovagnorio I. Gaetano Vinaccia's (1881-1971) Theoretical Work on the Relationship between Microclimate and Urban Design. Sustainability [Internet]. 2015;7(4):4448-73. Available from: http://www.mdpi.com/2071-1050/7/4/4448/htm

[71] Rudosfky B. Architecture without Architects. New York, USA: MOMA: New York; 1964.

[72] Olgyay V. Design with Climate: Bioclimatic Approach to Architectural Regionalism. Princeton, USA: Princeton University Press; 1963.

[73] BIG. Hot to cold. An odyssey of architectural adaptation. Berlin, Germany: Taschen; 2015. 712 p.

[74] Croset PA, Milesi S. A Bercy e Villejuif: due quartieri parigini a confronto. Casabella. 1994;(617).

7

Landscape Ecology Practices in Planning: Landscape Connectivity and Urban Networks

Ebru Ersoy

Abstract

The increasing need to conserve the nature and biodiversity and to maintain human well-being has motivated landscape planners and researchers to seek different planning approaches in urban environments. In this context, different approaches to planning urban networks have been developed to promote the sustainable use and functioning of landscapes, to conserve the nature and species, and increase its use and enjoyment by people [1, 2]. In principle, these approaches have been founded on the conservation of natural areas/biodiversity and with a consensus on their benefits to nature, biodiversity and people [3–5]. However, they generally differ from each other with respect to their expected aims, and ecological and/or social functions [6]. Therefore, by examining different planning approaches to networks, this chapter clarifies what is meant by these concepts and approaches in the literature.

Keywords: landscape planning, landscape ecology, landscape connectivity, sustainable landscapes, urban networks

1. Introduction

Landscape fragmentation affects habitats and wildlife and causes loss of connectivity [7–9]. The detrimental effects of fragmentation can be avoided or minimised by the creation of new habitats and/or the protection of existing habitats by ensuring more connected habitat patches (or the networks of habitats/green and open spaces). Hence, the growing awareness of the need for connected habitats/green and open spaces was reflected in planning approaches such as greenbelts and linked park systems, greenways, ecological networks, green networks and green infrastructure (GI).

These approaches have their own planning aims and strategies, in particular in the early stages of their development. But they have become closer with regard to their common concerns and the underlying concept of landscape connectivity to identify their spatial configuration. In the context of this chapter, the theoretical and scientific background of different network approaches has been reviewed.

The scope of early landscape planning approaches to networks was limited by their foci, where the spatial planning of nature and human dimensions has been treated as separate sectors. However, nature conservation and landscape planning practices have started to evolve into more integrated and multidisciplinary approaches, which are centered on the concepts of sustainability and multifunctionality [6, 10, 11].

A more recent planning approach is GI and it is thought to be able to ensure the multifunctionality of different land uses and provide sustainable benefits to nature, biodiversity and people from available land in and around urban environments. However, there is still need for a more robust decision-making structure and feasible planning approach as well as a measurable and traceable tool to planning GI in order to achieve these goals.

This chapter reviews and discusses the literature on landscape ecology applications in landscape planning with an emphasis on sustainability in urban environments. The chapter starts with urbanisation as an issue and explains how it leads to fragmentation and the loss of connectivity. After giving a brief overview of issues related to fragmentation, it moves on to the relationship between urbanisation and sustainability. Thereafter, it introduces different ways in which networks have been developed in ecology and planning to mitigate the adverse effects of fragmentation by enhancing landscape connectivity.

2. Urbanisation and sustainability

2.1. Urbanisation, fragmentation and connectivity

Urbanisation can be defined as a dynamic process, where the land is mainly modified with an extension of the urbanised area and/or increased population [12–14]. While different social processes are regarded as the main drivers of urbanisation (i.e., population growth and employment opportunities), the process of urbanisation itself affects social, economic and political life [14, 15].

In general, urbanisation is thought to have adverse effects on the nature, biodiversity, the quality of life of people as well as the functioning of local and global ecosystems [16]. Since the built-up areas cover a large proportion of the land surface in urban environments, they are generally thought to be more disturbed and degraded compared to rural areas [16, 17].

However, it has been claimed that the process of urbanisation may also provide favourable conditions for biodiversity as it creates and supports a variety of species because of the diversity of habitats included in urban environments [12, 18, 19]. This is exemplified in the work undertaken by Gaston et al. [20] who demonstrated that domestic gardens in Sheffield contain a large amount of biodiversity. Also, Savard et al. [21] drew our attention to some other benefits of urban ecosystems to people, species and the other aspects of biodiversity (e.g., population structure and genetic diversity). This is evident in the case of the cultivation of rare plants in urban areas, which may attract species that are dependent on those plants.

On the other hand, changes in the existing land uses/covers and fragmentation have assumed to be the most important environmental issues associated with the process of urbanisation [7, 22–24]. The term fragmentation reflects both a status and process. As a status, fragmentation can be defined as the degree of isolation of previously connected landscape components [25, 26]. As a process, it implies a dynamic process of structural and functional changes in a landscape where a continuous habitat type is split into separated patches with different sizes, shapes and functions [9, 10, 27].

Bennett [23, 24] summarises the major effects of fragmentation under structural changes in the landscapes and adverse effects on wildlife. With regard to its effects on landscape structure, fragmentation causes the loss and/or degradation of valuable habitats with an increasing isolation – or in other words the loss of connectivity [9, 23, 24, 28–30]. Recent evidence suggests that larger habitats can support a wide diversity of animal and plant species [31–33]. Accordingly, the loss or reduction of habitats also means a dramatic reduction in biodiversity, where some species become rare or extinct depending on their habitat requirements [9, 10, 32]. Therefore, the maintenance of connectivity has been regarded as a worldwide concern to mitigate the detrimental effects of fragmentation as well as the conservation of the nature and biodiversity.

The concept of connectivity stems from the relationships between the spatial structure and functioning of landscape and means "the degree to which a landscape facilitates or impedes movement of organisms among habitat patches" [4, 35, 36]. As one of the fundamental properties of landscapes, connectivity has been considered as "a measure of the ability of organisms to move among suitable habitat patches" [4, 30, 37].

According to another definition provided by Ahern [38], connectivity is "a spatial characteristic of systems which enables and supports the occurrence of specific processes and functions, through adjacency, proximity or functional linkage and connection". In this regard, the concept of connectivity encompasses the structural and functional aspects of a landscape. While structural connectivity refers to the degree to which habitat patches are physically/structurally linked to each other [23, 24, 39], functional connectivity denotes the measure of species' ability to move between habitat patches and does not necessarily require physical connections between habitat patches [19, 40, 41]. Functional connectivity, therefore, depends on the behavioural responses of organisms to the spatial structure of landscapes [39, 42, 43].

2.2. Urbanisation and sustainability

The relationship between urbanisation and sustainability largely depends on their dynamic interactions and interdependencies with environmental, societal and economic processes [44]. In urban areas, natural habitats and biodiversity have been subjected to intense human disturbances, and so urban environments and their surroundings have been the focus of conservation efforts [45]. In conjunction with the increased concerns for the nature and biodiversity, sustainability has become a central issue in urban areas, as a response to the growing concern for the quality of the natural environment as well as the social and economic life in the early nineteenth century [46, 47].

The concept of sustainable development is formally defined for the first time in the Brundtland Report (Our Common Future) as "development that meets the needs of the present without compromising the ability of future generations to meet their own needs" [48]. The concept of sustainable development seeks to achieve a dynamic and long-term balance between socioeconomic (e.g., well-being and equity of people) and environmental systems (e.g., protection and maintenance of the nature and biodiversity) [49]. As suggested by Selman [10], the landscape itself provides an arena in which this balance might be provided and maintained.

With regard to sustainability in landscapes, it is claimed that a generally accepted definition of landscape sustainability is lacking or it is generally defined in different contexts [49, 50]. While some researchers used the Brundtland Report as the source of definition with an emphasis on the maintenance of ecological integrity and basic human needs [27, 51], some highlighted the importance of natural capital and ecosystem services [52, 53], while others considered the localisation and self-regenerative capacity as the essential property of sustainability in landscapes [54]. However, in broad terms, landscape sustainability is defined as "the capacity of a landscape to consistently provide long-term, landscape-specific ecosystem services essential for maintaining and improving human wellbeing in a regional context and despite environmental and sociocultural changes" [49].

As with the three pillars of sustainability (environment, society and economy), landscape sustainability has been described on the basis of a variety of dimensions. Selman [50] draws attention to the five dimensions of landscape sustainability – i.e., environmental, economic, social, political and aesthetic sustainability. Likewise, Musacchio [55] describes six dimensions of landscape sustainability: environment, economy, equity, aesthetics, experience and ethics.

Selman [20] claims that environmental sustainability stems from landscape ecology as a response to the fragmentation of landscapes with an emphasis on the importance of landscape multi-functionality, ecosystem services and/or resilience. First of all, a sustainable landscape should maintain and improve landscape connectivity to facilitate species' life cycles as well as sustaining healthy and viable populations through a biodiverse network of habitats [50, 55, 56]. It should also be able to support other functions, provide a variety of ecosystem services to people, biodiversity and nature, besides its ability to achieve a state of relative stability [50, 57]. Another dimension of a sustainable landscape, the economic sustainability, draws attention to the importance of a "virtuous circle" in which the endogenous economic vitality of a local landscape maintains and supports environmental production practices (i.e., tourism, recreation, and the production of food and timber) as well as promoting landscape quality and the quality of life [10, 55, 58]. The social sustainability of a landscape, on the other hand, refers to opportunities for public participation in decision-making processes, inclusivity and equity in access, equal right to benefit from the use of resources, social inclusion and community cohesion [50, 58]. The political aspect of landscape sustainability focuses on an effective governance structure, where the planning, protection and management of different landscape elements are put in place on common grounds for both the public and private sectors [50]. Finally, the aesthetic sustainability of a landscape relates to the visual amenity and healthy functioning of the landscape [50, 55].

In view of all that has been mentioned so far, we can clearly see that a sustainable landscape is a multidimensional and dynamic system in which every component is crucial for providing multi-functionality, supporting the essential ecosystem services, benefiting the health and well-being of people as well as meeting the different needs of people in urban environments.

3. Landscape ecology and landscape planning

While a variety of definitions of landscape ecology have been proposed, one of the first people to use landscape ecology was Carl Troll [59], who integrated the different concepts of geography and ecology into a new inter-disciplinary research area [60, 61]. Defined as "the study of structure, function and change in a heterogeneous land area composed of interacting ecosystems" [62, 63], landscape ecology encompasses three main characteristics of landscapes: structure, function and change.

Landscape structure is the mosaic of different geographical units (ecosystems, etc.) and is characterised by the amount and occurrence of different units (composition) as well as their spatial distribution and arrangement in the landscape (configuration) [64–68]. Landscape function refers to the interaction between spatial components of landscapes (flow of matter, energy and organisms) and landscape change expresses alterations in the structure and functions of a landscape over time [27, 62, 69]. These three characteristics of landscapes are closely associated with each other and their relationships constitute the past, current and future landscapes [64, 70, 71].

Landscape ecology not only helps researchers to investigate the spatial structure and functions of changing landscapes, but it also can help to identify the origin of changes and the interactions between spatial structure, function and change of a landscape in order to find the most appropriate options for decision making [62, 63, 72–75]. Hence, it is clear that the science of landscape ecology provides valuable insights into how short and long term landscape planning processes can improve the quality of life and achieve sustainability in urban environments [74, 76].

According to a definition provided by the Landscape Institute [77], landscape planning is "the development and application of strategies, policies and plans to create successful environments, in both urban and rural settings, for the benefit of current and future generations". This definition refers to a formal process of decision making and technical/spatial planning activity built upon the assessment of physical, natural and cultural resources, where the main concern is the enhancement, restoration and/or creation of landscapes [10, 38, 78].

As mentioned earlier, urbanisation has been regarded as one of the main drivers of the change in landscape structure and function. As being one of the fundamental functions of a landscape, connectivity enables the movement of organisms through landscapes as well as sustaining other functional processes in a landscape [10]. Therefore, the crucial role of landscape connectivity for the conservation of nature and biodiversity has been emphasised by researchers to reduce the adverse effects of urbanisation, in particular the fragmentation and isolation of habitats [4, 32, 34, 69, 79, 80].

While planning is regarded as a key tool to deliver sustainable development [10], the need for multi-disciplinary and more integrated approaches to nature conservation and landscape planning has been highlighted [6]. Within this context, in order to mitigate the effects of fragmentation and to conserve nature and biodiversity in urban areas, different spatial planning approaches and strategies have been integrated into landscape planning and ecology, i.e., habitat creation or preservation that produces more connected patches or networks [7, 23, 26].

The growing recognition of connected systems was reflected in planning strategies such as greenbelts and linked park systems since the early nineteenth century. Thereafter a variety of approaches have been developed throughout the world with regard to the connected systems of green spaces in urban areas (i.e., ecological/green networks) [1, 2, 6, 38, 81, 82].

Even though each of these network approaches shares a great deal of common ground in terms of their main idea and structural properties, they typically differ from each other in their intended aims and functions that the networks will deliver [6]. While there has been a consensus on the main benefits of different network approaches for nature, biodiversity and people [4, 5, 83–85], there is a degree of uncertainty around the terminology on these approaches [3]. In this regard, it is necessary to clarify exactly what is meant by different network concepts and approaches in the literature. The next section therefore looks at the definition, the underlying rationale and the development of different networks being planned and managed in urban areas.

3.1. Linked park systems, green belts and greenways

The idea of greenways originates in the concept of parkways, the linear system of green and open spaces in urban areas, which was first developed by Frederick Law Olmsted in the nineteenth century [6, 86, 87]. Frederick Law Olmsted, the founder of the profession of landscape architecture in the USA, proposed two important plans for Brooklyn and Boston in order to connect urban parks and the surrounding areas as linear park systems. With a width between 65 and 150 m, these parkways aimed to deliver aesthetic and recreational functions for the benefit and use of people [6, 88, 89].

As the first greenway approach in the USA, the Boston Park System – or in other words the "Emerald Necklace" – was an attempt to integrate urban and suburban areas to increase the functioning of these areas [89–91]. On the other hand, in the same period in the UK, the concept of the "green belt" was first introduced by Ebenezer Howard, in his book *Garden Cities of To-Morrow* [92, 93]. Howard [94] claimed that if green and open spaces were located in close proximity to residential areas, they would contribute significantly to the physical and mental health of residents as well as their well-being.

The underlying idea of green belts was separating urban and rural environments from each other by designating some of the land around the inner cities as green to regulate urban sprawl and protect the countryside beyond the urban areas [6, 10, 93]. The fundamental difference between the traditional linked park systems and green belts is in their main functions. While the concept of parkways is built upon green corridors from and to the urban parks which are surrounded by trees (linking function), the concept of green belts is largely based on the idea of controlling urban growth by separating urban and rural areas with a buffer of undeveloped land (separating function) [6, 92]. Conversely, according to Kühn [95], in the future, green belts might behave as complementary zones between different urban areas by linking them in a polycentric city region [95]. Within this framework, being located in the urban fringe, green belts have the potential of providing a multifunctional and dynamic environment where there is a wide variety of low-density economic activities and a diversity of wildlife [96, 97].

At that time, the town planner Patrick Abercrombie developed another pioneering approach to planning and implementing a park system for Sheffield: the Sheffield Civic Survey and Development Plan. In this comprehensive city plan, Abercrombie [98] proposed a park system where all the individual open spaces (e.g., existing and new parks, playgrounds, accessible moorlands and waterwork properties) were linked to each other with tree-planted avenues.

According to Abercrombie [98], the success of a systematic provision of open spaces in urban areas is dependent on the area, use and distribution of open spaces. Accordingly, the following planning principles underpinned the Abercrombie's city plan:

- The area of different open spaces must be proportional to the whole extent of the city.
- The different uses and functions of open spaces should be determined by their user groups.
- Open spaces must be distributed throughout the city, in particular where it is appropriate and required. Therefore, planners should take into account the travelling distance to open spaces. Additionally, depending on the use of open spaces, certain types of parks should be located evenly throughout the city, whereas some of them must be placed in the city centre or distributed irregularly [98].

The underlying principle employed in Abercrombie's plan represented an emerging theoretical basis for the linked park systems, where it was suggested that all the green and open spaces should be located close to the centres of population it serves. Additionally, Winkler [99] claims that this strategic plan, grounded on an in-depth analysis process, has a crucial role in the development of Sheffield. Perhaps most importantly, Abercrombie's plan revealed the actual structure of Sheffield at that time, offering a complete framework for green and open spaces throughout the city and towards the Peak District National Park as well as making clear connections between green and open spaces and the centres of population [99].

Following these pioneering planning strategies, the concept of greenways has become a common landscape planning approach all over the world. Little [100] defined a greenway as the following:

- "A linear open space established along either a natural corridor, such as a riverfront, stream valley, or ridgeline, or overland along a railroad right-of way converted to recreational use, a canal, scenic road, or other route,
- Any natural or landscaped course for pedestrian or bicycle passage,
- An open-space connector linking parks, nature reserves, cultural features, or historic sites with each other and with populated areas,
- Locally, certain strips or linear parks designated as parkway or green belt".

Additionally, according to a definition provided by Ahern [3], greenways are the "networks of land that are planned, designed and managed for multiple purposes including ecological, recreational, cultural, aesthetic, or other purposes compatible with the concept of sustainable land use". Thereafter, a further definition is given by Ahern [38] who describes greenways as "the connected systems of protected lands that are managed for multiple uses including: nature protection, recreation, agriculture, and cultural landscape protection".

As shown in the abovementioned definitions, the focus of greenways has been moved from a single purpose planning approach to a multifunctional network approach, which is intended to assist key ecological functions as well as supporting public enjoyment and movement in urban environments [101]. In this regard, Ahern [91] claimed that the term greenway is a generic description of various strategic landscape planning approaches and plans which embodies a multitude of concepts with the main aim of ensuring multifunctionality in urban areas.

3.2. Ecological networks

Historically, the term "ecological" was inserted into the network approaches in the Netherlands with the ecological infrastructure concept [102]. As with greenways, the concept of ecological networks has been an attempt to integrate landscape ecology into landscape planning in order to protect nature and biodiversity, manage natural resources, and also to connect people with nature conservation [6]. Even though these terms have been used interchangeably, as pointed out by Ahern [91, 103], the term ecological networks is more common in Europe, whereas the term greenways is more common in the USA. Also, while greenways initially aimed to provide access to people between urban and rural green and open spaces in the USA, ecological networks in Europe stemmed from the need to conserve species and habitats [6].

Similar to greenways, a variety of definitions have been suggested for ecological networks in the literature. Bennett [104] defined ecological networks as "the coherent systems of natural or semi-natural landscape elements configured and managed with the objective of maintaining or restoring ecological functions as a means of conserving biodiversity, besides providing appropriate opportunities for the sustainable use of natural resources". According to a definition provided by Jongman and Pungetti [6], ecological networks are "the systems of nature reserves and their interconnections that make a fragmented natural system coherent, so as to support more biological diversity than in its non-connected form".

Although differences of definitions exist, there appears to be some agreement that the concept of ecological networks is founded on the conservation of natural areas and biodiversity as well as the enhancement of the functioning of ecosystems [2, 105, 106]. In addition to these, Ignatieva et al. [107] claim that urban ecological networks are one of the most effective tools for providing physical, visual and ecological connectivity between urban areas and surrounding natural areas. Hence, the development and the integration of ecological networks into the planning system have been regarded as the spatial expression of the idea of landscape connectivity in planning activities [6].

In general terms, spatially, the structural elements of ecological networks are composed of core areas, which are usually protected by surrounding buffer zones and are linked to each other with linear corridors [1, 78, 108, 109].

As claimed by Jongman [88], core areas are generally defined on the basis of traditional nature conservation practices as the natural and seminatural areas of conservation concern or the ecologically important areas with high nature value. Thus, the primary functions of core areas are thought to be the conservation of nature and biodiversity by meeting the ecological requirements of species or ecosystems [78].

The main purpose of corridors is to enable dispersal and migration of animal and plant species by providing functional connections between core areas (e.g., ecosystems or habitats for species) [1, 78, 108]. With regards to their intended ecological functions, Bouwma et al. [109] emphasised the crucial role of the spatial arrangement, internal structure and management of corridors, where the more complex corridors can provide multiple functions for different animal and plant species. Within ecological networks, three types of corridors are defined on the basis of their spatial structures and they are landscape, linear and stepping stone corridors [1, 78]. While landscape corridors can be in various forms of linked landscape matrices, linear corridors are composed of linear landscape elements such as rivers or forest strips. Conversely, stepping stone corridors are composed of a range of small habitat patches within the landscape matrix.

Buffer zones, on the other hand, prohibit the damaging effects from external influences and maintain landscape processes within core areas and corridors by creating environmental gradients around these [78, 88, 110]. Finally, another spatial element of ecological networks mentioned by Bouwma et al. [109] includes sustainable use areas which refer to the exploitation of opportunities within the landscape mosaic for the maintenance of ecosystem services and sustainable use of natural resources [1]. Also, more intensive human uses are allowed in buffer zones and sustainable use areas only if these activities support the maintenance of ecosystem services and sustainability [78, 110].

The common goals of ecological networks are to maintain the functioning of ecosystems and to promote the sustainable use of natural resources by assisting policy sectors [6, 81]. Within this framework, ecological networks have been considered one of the most important landscape planning approaches to address issues associated with human-induced habitat depletion. The concept of ecological networks has therefore attracted the attention of conservationists and planners in Europe [111, 112]. In Europe, many international initiatives and strategies for ecological networks have been developed [105, 113]. For example, the Pan-European Ecological Network (PEEN) is thought to be one of the most ambitious international ecological network programmes. The aims of the PEEN programme are to ensure the following:

- the conservation of a full range of ecosystems, habitats, species and landscapes of European importance,
- the maintenance of the habitats that are large enough to conserve animal and plant species,
- the promotion of sufficient opportunities for species to disperse and migrate,
- the restoration of the damaged parts of the key environmental systems, and
- the prevention of potential threats on key environmental systems [10–114].

Overall, the ecological network approach has been regarded as an important tool to maintain some level of ecological structure and function in urban areas, since they are thought to provide habitats and ecological connectivity for species and to conserve the wildlife [6]. But, as indicated by Andrian [115], the emphasis of the wildlife and nature conservation has been a major driver for the development of urban ecological networks and there is still a need for integrating social and cultural values into ecological networks.

3.3. Green networks

The green network concept has been inserted into urban planning practices, principally based on the idea of ecological networks [116]. Accordingly, the ecological and green network concepts have been used synonymously. However, the transition from ecological networks to green networks has created a noticeable shift in the spatial planning of nature and human dimensions. In other words, while the focus of the ecological network concept was on the conservation of species and habitats, the concept of green networks has brought the needs of species and humans together under the same roof [6, 82]. Moreover, the concept of green networks recognises the crucial role of green and open spaces and the linkages between them to support and improve sustainable development and also to enhance the functioning of urban environments [82]. Here, it is also important to note that the multiple benefits (social, economic, health and environmental) of urban green and open spaces have already been recognised by researchers, planners and decision-makers [84, 117–121].

Barker [122] defined green networks as "natural, or permanently vegetated, physically connected spaces situated in areas otherwise built-up or used for intensive agriculture, industrial purposes or other intrusive human activities". Additionally, the term green network was used by Bennett [104] to refer to a "spatial planning tool for the purpose of balancing and integrating land uses". Thus, the concept of green networks has been seen as a multifunctional urban planning approach, in which the value and importance of natural, seminatural habitats and human-dominated habitats (e.g., urban green and open spaces) are appreciated to deliver benefits both for people and the environment.

Regarding the requirements of people in urban environments, a recent study by Scotland and Northern Ireland for Environmental Research claimed that green networks are capable of increasing the number of people visiting urban green spaces and the countryside by providing a safe environment for people to move across [85, 123]. Furthermore, the green network approach goes beyond the limited vision of developing individual green spaces in urban areas just for recreational and visual purposes and focuses on the functionally connected systems of formal and informal green and open spaces [122, 124].

According to Forest Research [82], the concept of green networks takes into account multiple functions offered by green spaces as well as their ability to support the movement of people and species by the interconnections between them. Within this framework, the differences between individual green and open spaces and a green network have been explained according to their functions and spatial configurations. While green spaces refer to publicly accessible individual green areas in urban environments, green networks reflect a strategically identified and functional system of green spaces for the benefit of people, habitats and biodiversity [82, 125].

As well as these important features of green networks, Barker [122] suggested that one of the major benefits of green networks is their ability to provide connections between urban and rural landscapes. Therefore, green networks are said to be able to fulfil the requirements of wildlife, support ecological processes and meet the recreational, visual and social needs of people. In most countries, even though green networks have been primarily developed for their benefits to nature and biodiversity, they also serve multiple uses and functions such as meeting the ecological requirements of species and providing recreational facilities to people [122]. For example, in Sheffield, reasons for conserving and improving a green network for people and wildlife are defined as the following:

- to increase and support biodiversity in Sheffield and the surrounding areas,
- to allow the dispersal and genetic exchange of species throughout the city,
- to reduce the adverse effects of fragmentation and isolation,
- to control and support a sustainable drainage system,
- to encourage the movement of people by increasing the access to open and green spaces, and countryside,
- to improve the well-being and health of people, and
- to improve the general character of the city as an attractive and healthy place [126].

From a theoretical point of view, it is obvious that the intended functions of green networks are broadly compatible with the main functions of ecological networks, which aim to support and enhance the movement of animal and plant species. Besides maintaining and enhancing urban biodiversity and nature, the green network approach also provides appropriate opportunities for the sustainable use of natural resources, and so is regarded as one of the fundamental components of a more sustainable urban environment [82]. To conclude, the green network approach has been inserted into the planning and management strategies as a broad concept to achieve multifunctionality for biodiversity and people in urban areas.

3.4. Green Infrastructure

As a more recent approach, the GI concept takes its theoretical and conceptual background from the abovementioned network approaches to provide multiple benefits for biodiversity, nature and people within an urban environment [127, 128]. For this reason, we can claim that the GI concept is not a new idea in landscape planning and management [129]. Accordingly, it can be suggested that the concept of GI is grounded on the recognition of the crucial role of green networks in the wider landscape to provide essential services, functions and resources. In this context, Rouse and Bunster [130] claim that the previous plans of green and open spaces (e.g., greenways) have been increasingly adapted as GI plans to provide environmental, economic and social benefits in urban environments.

GI is defined by Benedict and McMahon [127] as "an interconnected network of waterways, wetlands, woodlands, wildlife habitats, and other natural areas; greenways, parks and other conservation lands; working farms, ranches and forests; and wilderness and other open spaces that support native species, maintain natural ecological processes, sustain air and water resources and contribute to the health and quality of life for communities and people". In addition, according to Natural England [131], GI is "the network of multifunctional open spaces, waterways, trees and woodlands, parklands and open countryside within and between our cities, towns and villages". In an urban context, the Natural Environment White Paper defined GI as "the living network of green spaces, water and other environmental features in both urban and rural areas. It is often used in an urban context to cover benefits provided by trees, parks, gardens, road verges, allotments, cemeteries, woodlands, rivers and wetlands" [132]. Furthermore, Natural England [133] suggested that GI is "a strategically planned and delivered network comprising the broadest range of high quality green spaces and other environmental features".

The careful wording of these definitions includes three important ideas at the heart of the GI concept: connectivity in the form of networks, multifunctionality and green components [127, 129, 134]. In spite of the emphasis on the term "green", it is quite important to note that GI also includes the features of blue infrastructure, such as river systems, other water features and coastal environments [127, 131]. As mentioned earlier, connectivity refers to the functional linkages in a landscape for the movement of animals, plants and/or people as well as the flows of materials, nutrients and energy [30, 35, 41]. Accordingly, supporting and enhancing connectivity between (habitat) patches is an important issue for biodiversity and nature conservation, and also to support human well-being and health in GI planning [127, 134].

Moreover, taking into consideration the abovementioned definitions, it is obvious that the concept of multifunctionality is the core idea of the GI concept, since the ability of a landscape to deliver multiple benefits and functions for wildlife, nature and people has been widely recognised by decision-makers, planners and managers. Multifunctionality refers to "the potential for GI to have a range of functions to deliver a broad range of ecosystem services"

[135]. The key benefits of a GI approach are summarised by Forest Research [136] as the following:

- mitigating and adapting to the effects of climate change,
- supporting and promoting the health and well-being of people,
- supporting economic growth and investment,
- regeneration of previously developed, derelict, underused and neglected (brown fields) lands,
- protecting, supporting and improving wildlife and habitats, and
- enhancing social inclusion and creating community cohesion.

The Landscape Institute [137] claims that a strategically planned and managed GI approach may provide enhanced multifunctions in comparison with the sum of individual green and open spaces in an urban area. The concept of multifunctionality can be fitted into the planning of individual green and open spaces as well as routes but we can achieve a fully multifunctional GI network when these individual sites and their connections are taken together [10]. In this respect, it is important to note that multifunctionality in a landscape is characterised by a high level of complexity, where different functions occur at the same time and interact with each other [138].

Consisting of natural, seminatural and man-made ecological systems altogether in a system, a GI forms a multifunctional network within and around urban areas [139]. Hence, the planning and management of a GI approach should take into account its capacity to deliver multiple ecological services, address the requirements of people and enhance the spatial character and quality of landscapes in urban environments [96, 131, 133]. Accordingly, a GI approach also requires involvement of a variety of stakeholders (e.g., planning authorities, policy makers, conservationists and the general public) in order to meet its intended functions and benefits that we expect them to provide [5, 140, 141].

In brief, the GI is a more promising and comprehensive planning approach to develop a coherent system of green and open spaces which serve multiple purposes and provide multifunctionality in urban areas [142].

4. Discussions and conclusions

The objective of this chapter was to provide a deeper understanding of the context and evolution of different approaches to planning and designing urban networks. While, during their early stages, different approaches had their own planning aims and strategies to define networks spatially, subsequently, their general concerns about nature, wildlife and people have become more aligned [6].

In early network planning practices, although natural and seminatural habitats were connected to each other for the benefit and use of wildlife, the linkages between urban green and open spaces were more concerned with people's use and enjoyment of nature. However, in urban environments, it is hard to develop a network which focuses only on the conservation of nature and biodiversity or the benefit of people [3]. In many cases it is not appropriate to apply such an approach, since we cannot ignore the interactions between nature and people in urban environments. In this context, there has been a shift from single purpose planning approaches to more comprehensive and integrative planning approaches in order to deliver multifunctionality in urban environments [10, 80].

All network approaches recognise the importance of functional connections for biodiversity and people in an increasingly fragmented urban environment. Accordingly, the common characteristics of the different network approaches are their spatial configuration and focus on connectivity. With regard to their spatial configuration in landscapes, all networks benefit from a linear structure in which different habitats and green and open spaces are included and connected. In relation to that, there is evidence for the benefits of networks to wildlife and biodiversity and people. The wider benefits to wildlife and biodiversity include facilitating the dispersal, genetic exchange and variability of many animal and plant species; increasing species' resilience to the environmental changes, predators and human disturbances and supporting the essential ecosystem services [2, 4, 83]. The benefits of networks to people, on the other hand, include supporting the health and well-being of people and enhancing community spirit [5, 84, 85, 136].

However, our understanding of the underlying science and the ways of planning, designing and managing networks in urban landscapes is still developing. Accordingly, one of the most important obstacles to enhance connectivity, maintain biodiversity and support human well-being through the planning of networks has been the gap between their intended aims and actual outcomes in urban environments.

A comprehensive network planning approach requires the following considerations in order to achieve sustainability and multifunctionality in urban environments. Sustainability is a natural characteristic of any planning activity. Accordingly, the spatial planning of networks requires the investigation and integration of ecological, societal and economic aspects to provide multiple benefits to wildlife, nature and people. Also, detailed research is required to explore how differing land use/cover morphologies within the wider landscape matrix would support or detract from their expected functions. In addition, after the determination of areas for different functions and/or multifunctionality, their applicability should be evaluated using different tools, such as the use of multicriteria analysis as well as defining opportunities and constraints for the planning decisions. Moreover, the planning strategies for networks require cooperation between the local and regional authorities to provide and support connectivity at landscape and regional levels. It is also important to identify and cooperate with stakeholders including public/private sectors and organisations to provide and support multifunctionality and sustainability in urban environments. Finally, it is a necessity to monitor the results of networks in order to measure and ensure the success of the network plans as well as identifying any changes and modifications to these plans.

Acknowledgements

This chapter is built upon as part of my PhD thesis. Hence, I gratefully acknowledge the invaluable guidance and support provided by Dr Anna Jorgensen (University of Sheffield, Department of Landscape) and Prof Philip H. Warren (University of Sheffield, Department of Animal and Plant Sciences) during my PhD research in the Department of Landscape, University of Sheffield.

Author details

Ebru Ersoy*

Address all correspondence to: ebruersy@gmail.com

Department of Landscape Architecture, Faculty of Agriculture, Adnan Menderes University, Aydin, Turkey

References

[1] Bennett G., Mulongoy K.J. Review of Experience with Ecological Networks, Corridors, and Buffer Zones. CBD Technical Series No 23. 2006. 100p.

[2] Lawton J.H., Brotherton P.N.M., Brown V.K., Elphick C., Fitter A.H., Forshaw J., Haddow R.W., Hilborne S., Leafe R.N., Mace G.M., Southgate M.P., Sutherland W.J., Tew T.E., Varley J., Wynne G.R. Making Space for Nature: A Review of England's Wildlife Sites and Ecological Network. Report to Defra. 2010. 107p.

[3] Ahern J. Greenways as a planning strategy. Landscape and Urban Planning. 1995; 33.1: 131–155.

[4] Taylor P.D., Fahrig L., With K.A. Landscape connectivity: a return to the basics. 2006: 14–29. In: Crooks, K.R., Sanjayan, M. (Eds.). Connectivity Conservation. Cambridge, UK: Cambridge University Press. 2006. 732p.

[5] Horwood K. Green infrastructure: reconciling urban green space and regional economic development: lessons learnt from experience in England's north-west region. Local Environment. 2011; 16.10: 963–975.

[6] Jongman R.H.G., Pungetti G. (Eds.). Ecological Networks and Greenways Concept, Design and Implementation. Cambridge Studies in Landscape Ecology. Cambridge, UK: Cambridge University Press. 2004. 368p.

[7] Saunders D.A., Hobbs R.J., Margules C.R. Biological consequences of ecosystem fragmentation: a review. Conservation Biology. 1991; 5.1: 18–32.

[8] Fahrig L. Effects of habitat fragmentation on biodiversity. Annual Review of Ecology, Evolution and Systematics. 2003; 34.8: 487–515.

[9] Lindenmayer D.B., Fisher J. Habitat Fragmentation and Landscape Change: An Ecological and Conservation Synthesis. Covelo, California: Island Press. 2006. 352p.

[10] Selman P. Planning at the Landscape Scale. Oxon: Routledge. 2006. 225p.

[11] Selman P. Centenary paper: Landscape planning – preservation, conservation and sustainable development. Town Planning Review. 2010; 81.4: 381–906.

[12] Niemelä J. Is there a need for a theory of urban ecology? Urban Ecosystems. 1999; 3.1: 57–65.

[13] Niemelä J., Kotze D.J., Yli-Pelkonen V. Comparative Urban Ecology: Challenges and Possibilities. 2009: 9–24. In: McDonnell, M.J., Hahs, A.K., Breuste, J. (Eds.). Ecology of Cities and Towns: A Comparative Approach. Oxford: Cambridge University Press. 2009. 746p.

[14] Dawson R.J., Wyckmans A., Heidrich O., Köhler J., Dobson S., Feliu E. Understanding Cities: Advances in Integrated Assessment of Urban Sustainability. Newcastle, UK: Centre for Earth Systems Engineering Research (CESER). 2014. 232p.

[15] Macionis J.J., Parrillo V.N. Cities and Urban Life. 3rd ed. Upper Saddle River, NJ: Pearson Education. 2004. 395p.

[16] Alberti M. The effects of urban patterns on ecosystem function. International Regional Science Review. 2005; 28.2: 168–192.

[17] Pickett S.T.A., Cadenasso M.L., Grove J.M., Nilon C.H., Pouyat R.V., Zipperer W.C., Costanza R. Urban ecological systems: linking terrestrial ecological, physical, and socioeconomic components of metropolitan areas. Annual Review of Ecological Systems. 2001; 32.1: 127–157.

[18] Gilbert O.L. The Ecology of Urban Habitats. London: Chapham & Hall. 1989. 369p.

[19] Andersson E. Urban Landscapes and Sustainable Cities. Ecology and Society. 2006; 11.1: 34. [Internet]. Available from: http://www.ecologyandsociety.org/vol11/iss1/art34/ [accessed: 2016-01-14].

[20] Gaston K.J., Warren P.H., Thompson K., Smith R.M. Urban domestic gardens (IV): the extent of the resource and its associated features. Biodiversity and Conservation. 2005; 14: 3327–3349.

[21] Savard J.P.L., Clergeau P., Mennechez G. Biodiversity concepts and urban ecosystems. Landscape and Urban Planning. 2000; 48.3: 131–142.

[22] Turner M.G., Gardner R.H., O'Neill R.V. Landscape Ecology in Theory and Practice: Pattern and Process. New York: Springer-Verlag. 2001. 404p.

[23] Bennett A.F. Linkages in the Landscape: The Role of Corridors and Connectivity in Wildlife Conservation. International Union for Conservation of Nature and Natural Resources: Gland, Switzerland and Cambridge, UK. 1998. 262p.

[24] Bennett A.F. Linkages in the Landscape: The Role of Corridors and Connectivity in Wildlife Conservation. International Union for Conservation of Nature and Natural Resources. 2nd ed.: Gland, Switzerland and Cambridge, UK. 2003. 254p.

[25] Franklin A.B., Noon B.R., George T.L. What is habitat fragmentation? Studies in Avian Biology. 2002; 25: 20–29.

[26] Bennett A.F., Saunders D.A. Habitat Fragmentation and Landscape Change. 2010: 88–106. In: Sodhi, N.S., Ehrlich, P.R. (Eds.). Conservation Biology for All. Oxford: Oxford University Press. 2010. 352p.

[27] Forman R.T.T. Land Mosaics: The Ecology of Landscapes and Regions. Cambridge: Cambridge University Press. 1995. 656p.

[28] Farina A. Principles and Methods in Landscape Ecology. Chapman & Hall: London. 1998. 235p.

[29] With K.A. Landscape Connectivity and Metapopulation Dynamics. 2002: 208–227. In: Gergel, S.E., Turner, M.G. (Eds.). Learning Landscape Ecology. A Practical Guide to Concepts and Techniques. New York, NY: Springer-Verlag. 2002. 316p.

[30] Hilty J.A., Lidicker W.Z., Merelender A.M. Corridor Ecology: The Science and Practice of Connectivity for Biodiversity Conservation. Washington DC: Island Press. 2006. 344p.

[31] Donovan T.M., Thompson F.R., Faaborg J., Probst J.R. Reproductive success of migratory birds in habitat sources and sinks. Conservation Biology. 1995; 9.6: 1380–1395.

[32] Farina A. Principles and Methods of Landscape Ecology. 2nd ed. Dordrecht, Netherlands: Springer. 2006. 412p.

[33] Debinski D.M., Holt R.D. A survey and overview of habitat fragmentation experiments. Conservation Biology. 2000; 14.2: 342–355.

[34] Noss R.F. Landscape Connectivity: Different Functions at Different Scales. 1991: 27–39. In: Hudson, W.E. (Ed.). Landscape Linkages and Biodiversity. Covelo, California: Island Press. 1991. 222p.

[35] Taylor P.D., Fahrig L., Henein K., Merriam G. Connectivity is a vital element of landscape structure. Oikos. 1993; 68.3: 571–573.

[36] Tischendorf L., Fahrig L. How should we measure landscape connectivity? Landscape Ecology. 2000; 15.7: 633–641.

[37] With K.A., Gardner R.H., Turner M.G. Landscape connectivity and population distributions in heterogeneous environments. Oikos. 1997; 78.1: 151–169.

[38] Ahern J. Greenways in the USA: Theory, Trends and Prospects. 2003: 34–55. In: Jongman, R.H.G., Pungetti, G. (Eds.). Ecological Networks and Greenways: Concept, Design and Implementation. United Kingdom: Cambridge University Press. 2003. 368p.

[39] Watts, K., Handley, P., Scholefield, P., Norton, L. Habitat Connectivity – Developing an Indicator for UK and Country Level Reporting. Phase 1 Pilot Study Contract Report to Defra. Forest Research and Centre for Ecology and Hydrology. Defra. [Internet]. 2008. Available from: http://nora.nerc.ac.uk/6875/1/Connectivity_indicator_0388_final_report_12Sept08_FINAL.pdf [accessed: 2016-01-14].

[40] Baudry J., Merriam H.G. Connectivity and Connectedness: Functional versus Structural Patterns in Landscapes. 1988: 23–38. In: Schreiber, K.F. (Ed.). Connectivity in Landscape Ecology (Proceedings of the 2nd International Seminar of the International Association for Landscape Ecology). Münstersche Geographische Arbeiten. Germany. 1988.

[41] Burel F., Baudry J. Landscape Ecology. Concepts, Methods, and Application. USA: Science Publishers, Inc. Enfield (NH). 2003. 378p.

[42] Collinge S.K. Ecology of Fragmented Landscapes. Baltimore, Maryland: Johns Hopkins University Press. 2009. 360p.

[43] Meiklejohn K., Ament R., Tabor, G. Habitat Corridors and Landscape Connectivity: Clarifying the Terminology. Center for Large Landscape Conservation. 2009. Available from: http://www.wildlandsnetwork.org/sites/default/files/terminology %20CLLC.pdf [accessed: 2016-01-16].

[44] Antrop M. Sustainable landscapes: contradiction, fiction or utopia? Landscape and Urban Planning. 2006; 75.3: 187–197.

[45] Gill S., Handley J., Ennos R., Pauleit S. Adapting cities for climate change: the role of the green infrastructure. Built Environment. 2007; 33: 115–133.

[46] Du Pisani J.A. Sustainable development – historical roots of the concept. Environmental Sciences. 2006; 3.2: 83–96.

[47] Kuhlman T., Farrington J. What is sustainability? Sustainability. 2010; 2.11: 3436–3448.

[48] World Commission on Environment and Development (WCED). Our Common Future, New York: Oxford University Press, 1987.

[49] Wu J. Landscape sustainability science: ecosystem services and human well-being in changing landscapes. Landscape Ecology. 2013; 28.6: 999–1023.

[50] Selman P. What do we mean by "sustainable landscape"? Sustainability: Science, Practice, and Policy. 2008; 4: 23–28.

[51] Turner M.G., Donato D.C., Romme W.H. Consequences of spatial heterogeneity for ecosystem services in changing forest landscapes: priorities for future research. Landscape Ecology. 2013; 28.6: 1081–1097.

[52] Haines-Young R. Sustainable development and sustainable landscapes: defining a new paradigm for landscape ecology. Fennia. 2000; 178.1: 7–14.

[53] Potschin M., Haines-Young R. "Rio+ 10", sustainability science and landscape ecology. Landscape and Urban Planning. 2006; 75.3: 162–174.

[54] Selman P. Landscape and Sustainability at the National and Regional Scales. 2007: 104–117. In: Benson, J.F., Roe, M. (Eds.) Landscape and Sustainability. 2nd ed. Taylor & Francis. 2007. 336p.

[55] Musacchio L.R. The scientific basis for the design of landscape sustainability: a conceptual framework for translational landscape research and practice of designed landscapes and the six Es of landscape sustainability. Landscape Ecology. 2009; 24.8: 993–1013.

[56] Termorshuizen J.W., Opdam P., Van den Brink A. Incorporating ecological sustainability into landscape planning. Landscape and Urban Planning. 2007; 79.3: 374–384.

[57] Harris, J.M. Sustainability and Sustainable Development. International Society for Ecological Economics. [Internet]. 2003. Available from: http://isecoeco.org/pdf/susdev.pdf [accessed: 2016-01-11].

[58] Benson J.F., Roe M. (Eds) Landscape and Sustainability. 2nd ed. Taylor & Francis. 2007. 336p.

[59] Troll, C. Aerial photography and ecological studies of the earth. Zeitschraft der Gesellschaft fur Erdkunde Zu Berlin. 1939: 241–298.

[60] Naveh Z., Lieberman A.S. Landscape Ecology Theory and Application. New York, USA: Springer-Verlag. 1984. 360p.

[61] Turner M.G. Landscape ecology: what is the state of the science? Annual Review of Ecology, Evolution, and Systematics. 2005; 36.2005: 319–344.

[62] Forman R.T.T., Godron M. Landscape Ecology. New York: John Wiley & Sons. 1986. 619p.

[63] Golley F.B., Bellot J. Interactions of landscape ecology, planning and design. Landscape and Urban Planning. 1991; 21.1-2: 3–11.

[64] Turner M.G. Landscape ecology: the effect of pattern on process. Annual Review of Ecology and Systematics. 1989; 20: 171–197.

[65] Turner M.G., Gardner R.H. (Eds.). Quantitative Methods in Landscape Ecology: The Analysis and Interpretation of Landscape Heterogeneity. New York: Springer-Verlag. 1991. 536p.

[66] McGarigal, K., Marks, B.J. FRAGSTATS Spatial Pattern Analysis Program for Quantifying Landscape Structure. Version 2.0. Forest Science Department, Oregon State University, Corvallis. [Internet]. 1995. Available from: http://www.umass.edu/landeco/pubs/mcgarigal.marks.1995.pdf [accessed: 2015-12-17].

[67] Gergel S.E., Turner M.G. (Eds.). Learning Landscape Ecology. A Practical Guide to Concepts and Techniques. New York: Springer-Verlag. 2002. 316p.

[68] McGarigal, K. FRAGSTATS HELP. Version 4.2. University of Massachusetts, Amherst. [Internet]. 2015. Available from: http://www.umass.edu/landeco/research/fragstats/documents/fragstats.help.4.2.pdf [accessed: 2016-02-17].

[69] Farina A. Landscape Ecology in Action. Dordrecht. The Netherlands: Kluwer Academic Publishers. 2000. 317p.

[70] Van Dorp D., Opdam P.F.M. Effects of patch size, isolation and regional abundance on forest bird communities. Landscape Ecology. 1987; 1.1: 59–73.

[71] Fairclough G. Europe's Landscape: Archaeology, Sustainability and Agriculture. In: Fairclough, G. Rippon, S. (Eds.). Europe's Cultural Landscape: Archaeologists and the Management of Change, Europae Archaeologiae Consilium Occasional Paper 2, Belgium, pp. 1–12.

[72] Urban D.L., O'Neill R.V. Jr., Shugart H.H. Landscape ecology. BioScience. 1987; 37.2: 119–127.

[73] Dramstad W., Olson J., Forman R. Landscape Ecology Principles in Landscape Architecture and Landscape Planning. Washington DC, Island Press. 1996. 80p.

[74] Boothby J. An ecological focus for landscape planning. Landscape Research. 2000; 25.3: 281–289.

[75] Botequilha-Leitão A., Ahern J. Applying landscape ecological concepts and metrics in sustainable landscape planning. Landscape and Urban Planning. 2002; 59.2: 65–93.

[76] Weddle A.E. Landscape Techniques: Incorporating Techniques of Landscape Architecture. London: Heinemann. 1979. 265p.

[77] Landscape Institute. Landscape Architecture: Elements and Areas of Practice. An Educational Framework. [Internet]. 2012. Available from: http://landscapeinstitute.org/PDF/Contribute/A4_Elements_and_areas_of_practice_education_framework_Board_final_2012.pdf [accessed: 2016-02-17].

[78] Council of Europe. The European Landscape Convention. Strasbourg, Cedex, France. [Internet]. 2000. Available from: http://www.coe.int/en/web/landscape/home [accessed: 2016-02-17].

[79] Collinge S.K., Forman R.T.T. A conceptual model of land conversion processes: Predictions and evidence from a microlandscape experiment with grassland insects. Oikos. 1998; 82.5: 66–84.

[80] Noss R.F., Dobson A.P., Baldwin R., Beier P., Davis C.R., Dellasala D.A., Francis J., Locke H., Nowak K., Lopez R., Reining C., Trombulak S.C., Tabor G. Bolder thinking for conservation. Conservation Biology. 2012; 26.1: 1–4.

[81] Bennett, G., Wit, P. The Development and Application of Ecological Networks: A Review of Proposals, Plans and Programmes. AID Environment Advice and Research for Development and Environment and IUCN World Conservation Union: Gland, Switzerland. [Internet]. 2001. Available from: https://portals.iucn.org/library/efiles/documents/2001-042.pdf [accessed: 2016-02-17].

[82] Forest Research. Green Networks and People; A Review of Research and Practice in the Analysis and Planning of Multi-Functional Green Networks. Scottish Natural Heritage Commissioned Report No. 490. [Internet]. 2011. Available from: http://www.snh.org.uk/pdfs/publications/commissioned_reports/490.pdf [accessed: 2016-02-17].

[83] Crooks K.R., Sanjayan M. (Eds). Connectivity Conservation. Conservation Biology Book Series. Cambridge, UK: Cambridge University Press. 2006. 732p.

[84] Dunnett, N., Swanwick, C., Woolley, H. Improving Urban Parks, Play Areas and Green Spaces. Office of the Deputy Prime Minister. London. [Internet]. 2002. Available from: http://publiekeruimte.info/Data/Documents/e842aqrm/53/Improving-Urban-Parks.pdf [accessed: 2016-02-18].

[85] CABE. Urban Green Nation: Building the Evidence Base. CABE Publication. London: Commission for Architecture and the Built Environment. [Internet]. 2010. Available from: http://webarchive.nationalarchives.gov.uk/20110118095356/http://www.cabe.org.uk/files/urban-green-nation.pdf [accessed: 2016-02-18].

[86] Makhzoumi J., Pungetti G. Ecological Landscape Design and Planning: The Mediterranean Context. London: Taylor & Francis. 1999. 352p.

[87] Makhzoumi J., Pungetti G. Ecological Landscape Design and Planning: The Mediterranean Context. 2nd ed. London: Taylor & Francis e-Library. 2005.

[88] Jongman R.H.G. The Context and Concept of Ecological Networks. 2004: 7–33. In: Jongman, R.H.G., Pungetti, G. (Eds.). Ecological Networks and Greenways Concept, Design and Implementation. Cambridge Studies in Landscape Ecology. Cambridge, UK: Cambridge University Press. 2004. 368p.

[89] Fabos G.J. Greenway planning in the United States: its origins and recent case studies. Landscape and Urban Planning. 2004; 68.2: 321–342.

[90] Zube E.H. Greenways and the US National Park System. Landscape and Urban Planning. 1995; 33.1-3: 17–25.

[91] Ahern, J. Greenways as Strategic Landscape Planning: Theory and Application. Published Dissertation. Wageningen University: Wageningen, The Netherlands. [Internet]. 2002. 182p. Available from: https://works.bepress.com/ahern_jack/7/ [accessed: 2016-01-14].

[92] Ndubisi F. Ecological Planning: A Historical and Comparative Synthesis. Baltimore, Maryland: The Johns Hopkins University Press. 2002. 304p.

[93] Amati M. Green Belts: A Twentieth-century Planning Experiment. In: Amati M. (Eds). Urban Green Belts in the Twenty-first Century. Aldershot: Ashgate. 2008. 1–18.

[94] Howard E. Garden Cities of To-Morrow. London: S. Sonnenschein & Co., Ltd. 1902. 195p.

[95] Kühn M. Greenbelt and Green Heart: separating and integrating landscapes in European city regions. Landscape and Urban Planning. 2003; 64.1–2: 19–27.

[96] Gallent N., Shoard M., Andersson J., Oades R., Tudor C. England's Urban Fringes: multi-functionality and planning. Local Environment. 2004; 9:217–233.

[97] Hall M.C., Page S.J. The geography of tourism and recreation: Environment, place and space. Routledge. 2014. 456p.

[98] Abercrombie P. Sheffield: A Civic Survey and Suggestions towards a Development Plan. University of Liverpool. 1924. 85p.

[99] Winkler, A. Sheffield City Report. Case Report 45. (London: London School of Economics and Political Science. [Internet]. 2007. Available from: http://sticerd.lse.ac.uk/dps/case/cr/CASEreport45.pdf [accessed: 2016-01-11].

[100] Little C.E. Greenways for America (Creating the North American Landscape). Baltimore/London: The John Hopkins University Press. 1990. 288p.

[101] Mugavin D. Adelaide's Greenway: River Torrens Linear Park. Landscape and Urban Planning. 2004; 68.2-3: 223–240.

[102] Hailong L., Dihua L., Xili H. Review of ecological infrastructure: concept and development. City Planning Review. 2005; 29.9: 70–75.

[103] Ahern J. Theories, Methods and Strategies for Sustainable Landscape Planning. From Landscape Research to Landscape Planning. Aspects of Integration, Education and Application. Springer, Dordrecht, NL 2006: 119–131.

[104] Bennett, G. Integrating Biodiversity Conservation and Sustainable Use: Lessons Learned From Ecological Networks. IUCN. Gland, Switzerland and Cambridge, UK. [Internet]. 2004. Available from: https://portals.iucn.org/library/efiles/edocs/2004-002.pdf [accessed: 2016-01-19].

[105] Jongman R.H.G., Külvik M., Kristiansen I. European ecological networks and greenways. Landscape and Urban Planning. 2004; 68.2-3: 305–319.

[106] Opdam P., Steingröver E., Van Rooij S. Ecological networks: A spatial concept for multi-actor planning of sustainable landscapes. Landscape and Urban Planning. 2006; 75.3: 322–332.

[107] Ignatieva M., Stewart G.H., Meurk C. Planning and design of ecological networks in urban areas. Landscape and Ecological Engineering. 2011; 7.1:17–25.

[108] Bischoff, N.T, Jongman, R.H.G. Development of Rural Areas in Europe: The Claim for Nature. Netherlands Scientific Council for Government Policy. Preliminary and Background Studies, V79. SDU: The Hague. [Internet]. 1993. 206p. Available from: http://www.wrr.nl/fileadmin/nl/publicaties/DVD_WRR_publicaties_1972-2004/V079Development_of_Rural_Areas_in_Europe_.pdf [accessed: 2016-01-19].

[109] Bouwma I.M., Jongman, R.H.G., Butovsky, R.O. Indicative Map of the Pan-European Ecological Network for Central and Eastern Europe. 2002. 166p.

[110] Schlumprecht H. Mapping the European Green Belt. 2006: 147–159. In: Terry, A., Ullrich, K., Riecken, U. The Green Belt of Europe: From Vision to Reality. IUCN, Gland, Switzerland and Cambridge, UK. 2006. 224p.

[111] Jongman R.H., Kristiansen I. National and Regional Approaches for Ecological Networks in Europe. Nature and Environment, No. 110. Strasbourg: Council of Europe Publishing. 2001. 86p.

[112] Boitani L., Falcucci A., Maiorano L., Rondinini C. Ecological networks as conceptual frameworks or operational tools in conservation. Conservation Biology. 2007; 21.6: 1414–1422.

[113] Tillmann J.E. Habitat fragmentation and ecological networks in Europe. Ecological Perspectives for Science and Society. 2005; 14.2: 119–123.

[114] Gilbert K., Rientjes S., van't Erve S. Communicating the Pan European Ecological Network. An analysis of the implementation and communication processes for ecological networks in Europe. ECNC. 2005: 72–74.

[115] Andrian G. Joining Cultural and Natural Heritage along the Green Belt. The Green Belt of Europe: From Vision to Reality. 2006: 20–25. In: Terry, A., Ullrich, K., Riecken, U. The Green Belt of Europe: From Vision to Reality. IUCN, Gland, Switzerland and Cambridge, UK. 2006. 224p.

[116] Külvik, M., Suškevičs, M., Kreisman, K. Current Status of the Practical Implementation of Ecological Networks in Estonia. [Internet]. 2008. Available from: http://www.ecologicalnetworks.eu/documents/publications/ken/EstoniaKENWP2.pdf [accessed: 2016-01-19].

[117] Takano T., Nakamura K., Watanabe M. Urban residential environments and senior citizens' longevity in megacity areas: the importance of walkable green spaces. Journal of Epidemiology and Community Health. 2002; 56.12: 913–918.

[118] Woolley H. Urban Open Spaces. London: Spon Press. 2003. 208p.

[119] Groenewegen P.P., Van den Berg A.E., De Vries S., Verheij R.A., Vitamin G. Effects of green space on health, well-being, and social safety. BMC Public Health. 2006; 6.1: 149.

[120] Office of the Deputy Prime Minister (ODPM) and National Audit Office (NAO). Enhancing Urban Green Space. In: Office of the Deputy Prime Minister (Ed.). London: The Stationery Office. [Internet]. 2006. Available from: http://www.nao.org.uk/wp-content/uploads/2006/03/0506935.pdf [accessed: 2016-01-19].

[121] Barbosa O., Tratalos J., Armsworth P.R., Davies R.G., Fuller R.A., Johnson P., Gaston K.J. Who Benefits from Access to Green Space? A Case Study from Sheffield, UK. Landscape and Urban Planning. 2007; 83.2-3: 187–195.

[122] Barker, G. A Framework for the Future: Green Networks with Multiple Uses in and around Towns and Cities. English Nature Research Reports No. 256. London: English Nature. [Internet]. 1997. Available from: http://publications.naturalengland.org.uk/publication/77041 [accessed: 2016-01-13].

[123] SNIFFER. Urban Network for People and Biodiversity – Form and Function. Edinburgh: SNIFFER. [Internet]. 2008. Available from: http://www.sniffer.org.uk/files/3413/4183/8008/UEUB01_Final_report_e-version_May_2008.pdf [accessed: 2016-01-13].

[124] Tzoulas K., James P. Peoples' use of, and concerns about, green space networks: a case study of Birchwood, Warrington New Town, UK. Urban Forestry and Urban Greening. 2010; 9.2: 121–128.

[125] Moseley D., Marzano M., Chetcuti J., Watts K. Green networks for people: application of a functional approach to support the planning and management of greenspace. Landscape and Urban Planning. 2013; 116.2013: 1–12.

[126] Sheffield City Council (SCC). Sheffield Local Plan Background Reports 2013. Sheffield Local Plan (formerly Sheffield Development Framework), City Policies and Sites Document Green Environment Policy Background Report. [Internet]. 2013. Available from: https://www.sheffield.gov.uk/dms/scc/management/corporatecommunications/documents/planning/SDF/background-reports-2013/Green-Environment--Policies-G1---G4-/Green%20Environment%20%28Policies%20G1%20-%20G4%29.pdf [accessed: 2014-03-05].

[127] Benedict M.A., McMahon, E.T. Green Infrastructure: Smart Conservation for the 21st Century. Washington, D.C., Sprawl Watch Clearing House. [Internet]. 2002. Available from: http://www.sprawlwatch.org/greeninfrastructure.pdf [accessed: 2016-01-13].

[128] Mell, I.C. Green Infrastructure: Concepts and Planning. In: FORUM Ejournal. 2008; 8.1: 69–80. [Internet]. 2008. Available from: http://research.ncl.ac.uk/forum/v8i1/green%20infrastructure.pdf [accessed: 2015-12-09].

[129] Wright H. Understanding green infrastructure: the development of a contested concept in England. Local Environment. 2011; 16.10: 1003–1019.

[130] Rouse D.C., Bunster-Ossa, I.F. Green Infrastructure: A Landscape Approach (No. 571). [Internet]. 2013. Available from: https://www.planning.org/pas/reports/subscriber/archive/pdf/PAS_571.pdf [accessed: 2015-12-09].

[131] Natural England. Natural England's Policy Position on Housing Growth and Green Infrastructure: Pre-Scoping Paper on Principles Natural England Board. London: Natural England. [Internet]. 2007. Available from: http://www.rudi.net/files/101007-NEBP0728.pdf [accessed: 2015-12-05].

[132] DEFRA. The Natural Choice: Securing the Value of Nature. [Internet]. 2011. Available from: https://www.gov.uk/government/uploads/system/uploads/attachment_data/file/228842/8082.pdf [accessed: 2015-12-12].

[133] Natural England. Green Infrastructure: Mainstreaming the Concept – Understanding and Applying the Principles of Green. London: Natural England. [Internet]. 2012. Available from: http://publications.naturalengland.org.uk/file/275811. [accessed: 2015-12-12].

[134] Benedict M.A., McMahon E.T. Green Infrastructure: Linking Landscapes and Communities. Island Press: Washington. 2006. 320p.

[135] Natural England. Green Infrastructure Guidance. London: Natural England. [Online]. 2009. Available from: http://publications.naturalengland.org.uk/publication/35033 [accessed: 2015-12-12].

[136] Forest Research. Benefits of Green Infrastructure. Report by Forest Research. Forest Research, Farnham. [Internet]. 2010. Available from: http://www.forestry.gov.uk/pdf/urgp_benefits_of_green_infrastructure.pdf/$FILE/urgp_benefits_of_green_infrastructure.pdf [accessed: 2015-01-21].

[137] Landscape Institute. Green Infrastructure and the Value of Connected, Multifunctional Landscapes. Landscape Institute Position Statement. [Internet]. 2009. Available from: http://www.landscapeinstitute.co.uk/PDF/Contribute/GreenInfrastructureposition-statement13May09.pdf [accessed: 2015-01-21].

[138] Selman P. Planning for landscape multifunctionality. sustainability: science, practice and policy. Community Essay. 2009; 5.2: 45–52.

[139] Tzoulas K., Korpela K., Venn S., Yli-Pelkonen V., Kaz´mierczak, A., Niemelä, J., James, P. Promoting ecosystem and human health in urban areas using green infrastructure: a literature review. Landscape and Urban Planning. 2007; 81.3: 167–178.

[140] Naumann S., McKenna D., Kaphengst T., et al. Design Implementation and Cost Elements of Green Infrastructure Projects. Final Report. Brussels: European Commission. [Internet]. 2011. Available from: http://ec.europa.eu/environment/enveco/biodiversity/pdf/GI_DICE_FinalReport.pdf [accessed: 2015-01-19].

[141] Naumann S., Anzaldua G., Berry P., et al. Assessment of the Potential of Ecosystem-based Approaches to Climate Change Adaptation and Mitigation in Europe. Final Report to the European Commission, EnvironmentDG, Contract No. 70307/2010/580412/SER/B2. Brussels; European Commission. [Internet]. 2011. Available from: http://ec.europa.eu/environment/nature/climatechange/pdf/EbA_EBM_CC_FinalReport.pdf [accessed: 2015-01-19].

[142] Sandström U.G. Green infrastructure planning in urban Sweden. Planning Practice and Research. [Internet]. 2002; 17.4: 373–385.

8

Effects of Urbanization and the Sustainability of Marine Artisanal Fishing: A Study on Tropical Fishing Communities in Brazil

Simone F. Teixeira, Daniele Mariz,
Anna Carla F. F. de Souza and Susmara S. Campos

Abstract

Marine fishing occurs along the coast and oceanic islands within the Exclusive Economic Zone of Brazil and is practiced mainly in an industrial fashion in the southern and southeastern regions of the country as well as in an artisanal fashion in the northern and northeastern regions. Artisanal marine fishing is practiced by fishermen who use sailing rafts or mid-size motorboats in daily fishing activities or activities that surpass 20 days at sea. To face the ocean and extract sustenance and income, the majority of artisanal fishermen do not have advanced fishing technologies, but rather empirical knowledge passed from one generation to another, which has allowed fishermen to maintain their activities for hundreds of years. The shared knowledge with regard to fishing and gear as well as fishing territories and the discovery of new territories allows artisanal fishermen to maintain catches while resources have become scarce. However, different factors in urban areas have been contributing to changes in artisanal marine fishing, such as the facility to education and jobs in other sectors of the economy. The aim of the present study was to investigate the experiences of artisanal fishermen in traditional communities of northeastern Brazil in the occurrence of urbanization. Traditionally in artisanal fishing, the transmission of knowledge occurs in the family, generally from father to son. However, this traditional transmission of knowledge is being lost in urban fishing communities. The urban environment facilitates access to formal education and provides opportunities for both formal and informal jobs, leading to income in more attractive sectors to the sons of fishermen than the activity of fishing. This is caused by changes in schooling and has triggered the avoidance of youths with regard to fishing activities. Moreover, urban pressures, such as the loss of areas of embarkation and landing, further hinder the maintenance of fishing in such areas. Thus, issues related to urbanization have been changing the structure of fishing communities and compromising the maintenance and sustainability of marine artisanal fishing activities in urban areas.

Keywords: Fishermen, traditional knowledge, artisanal fisheries, urban pressures, sustainable fisheries

---

1. Introduction

The coast of Brazil is considered a national heritage in the constitution passed in 1988 (Article 225, Section 4) [1] and corresponds to the geographic space of the interaction among the air, ocean and land, including renewable and non-renewable resources as well as the terrestrial area that encompasses municipalities under the direct influence of phenomena that occur in this zone and a marine area that extends for 12 nautical miles (Decree 5300, Article 3, 2004) [2]. The Brazilian Exclusive Economic Zone (EEZ) starts at the end of this marine area and extends an additional 200 nautical miles, in which, according to Article 7 of Law n° 8617 [3], "Brazil has sovereign rights for the exploitation, use, conservation and management of living and non-living natural resources of the waters over the ocean floor and its subsoil as well as other activities directed at the exploration and use of this zone for economic purposes." The zone extends from the tropical region at the mouth of the Oiapoque River in the state of Amapá (north of the equator) to the temperate region at the mouth of the Chuí Stream in the state of Rio Grande do Sul (southern Brazil), spanning 17 states and more than 8500 km.

The EEZ is a marine area of approximately 3.6 million km^2. However, adding the approximately 900 thousand km^2 that Brazil has formally requested from the United Nations, the total will be approximately 4.5 million km^2, corresponding to 52% of the mainland area. Due to this oceanic territory, the strategic importance and riches of living and non-living resources, the Brazilian Navy denominates this area the "Blue Amazon" [4]. The EEZ of the northeastern region of the country (EEZ/NE) extends from the mouth of the Parnaíba River in the state of Piauí to the city of Salvador in the state of Bahia.

The coast of Brazil is on the Atlantic Plate being characterized by three well-developed provinces: the continental shelf, continental slope and continental rise [5]. The continental shelf of the EEZ/NE is divided into two distinct stretches: (1) from the mouth of the Parnaíba River in the state of Piauí to Ponta do Calcanhar in the state of Rio Grande do Norte; and (2) from Ponta do Calcanhar to the city of Salvador in the state of Bahia [6]. In the first stretch, the continental shelf has a mean width of 63 Km and various platforms on the outer portion, which are well individualized between 23 and 30 m, 40 and 50 m as well as 60 and 70 m [6]. After Ponta do Calcanhar to Belmonte in the state of Bahia, the continental shelf gradually becomes narrower, with a maximum width of 42 km, depths down to 60 m, a quite irregular topography and typical bio-constructional occurrences (beach rocks), which are parallel to the coast and more developed from Cape Calcanhar southwards [5]. The North Brazilian Chain and volcanic Fernando de Noronha Chain, which is a line of underwater mounts, are also found off the northeastern continental shelf; only the Fernando de Noronha archipelago and Rocas atoll reach the surface [5].

The surface currents that traverse the northeastern continental shelf originate from the South Equatorial Current, which runs westward and splits at Cape Calcanhar to form the North Brazil Current, which passes over the Equator, whereas the larger part, the Brazil Current, runs in the southerly direction [7–8]. The South Equatorial Current and its two offshoots (North Brazil Current and Brazil Current) are warm and oligotrophic [9], but small-scale, highly transitory topographic resurgent phenomena occur around the chains of the EEZ/NE [10], which enrich the surface with nutrients from deep waters and enhance both primary and secondary productivity [11].

Small estuaries lined by mangroves are found on the northeastern coast of Brazil [12] and only the Parnaíba and São Francisco Rivers have large enough mouths to provide nutrients for the outer continental shelf [9]. The lack of large rivers and the occurrence of warm waters have led to the formation of coral reefs that extend approximately 3000 km from the state of Maranhão to the south of the state of Bahia. These ecosystems are home to the greatest diversity of fish fauna in marine environments [13], which have considerable ecological, economic and social importance to the region [14], sheltering important fish stocks and contributing to the subsistence of traditional coastal communities [15].

In this extensive stretch of ocean in off northeastern Brazil, with its geological, oceanographic and biological diversity, an enormous contingent of fishermen that mainly practice artisanal fishing operations exploit living resources, facing the ocean on a daily basis to extract their sustenance. Fisheries in the northeastern region are characterized by a wide variety of species with low abundance [12], but high commercial value [16] and multi-species catches. These fisheries are essentially artisanal (96.3%) and target pelagic (anchovies, halfbeaks and sardines), demersal and benthic (spotted goatfish, parrotfishes, tomtate grunt, drums, mutton snapper and white grunt) fish species as well as crustaceans (lobsters and shrimps) and mollusks [12].

The technology is unsophisticated, landings are decentralized and there is a lack of technical assistance and infrastructure from production through to sales [16]. Besides these problems related directly to fisheries, fishermen are affected by anthropogenic pressure on ecosystems, live resources and fishing territories. Fishermen in urban areas also face the difficulty or even impossibility of passing down their traditional knowledge regarding the environment and the resources they exploit, as the urban setting offers more attractive opportunities to the younger generation, making youths disinterested in fishery activities.

The object of the present study was marine artisanal fishermen on the northeastern coast of Brazil, who traditionally pass on knowledge regarding their occupation orally within the nuclear family. As urban fishing communities have been experiencing the breakdown of this tradition due to opportunities offered to youths to seek other ways of life as well as urban pressures that further hinder the perpetuation of fishing practices in such areas, the aim of the present investigation was to analyze the effects of urbanization on the main traditional fishing communities in northeastern Brazil (**Figure 1**) that can compromise the continuity and sustainability of marine artisanal fishing.

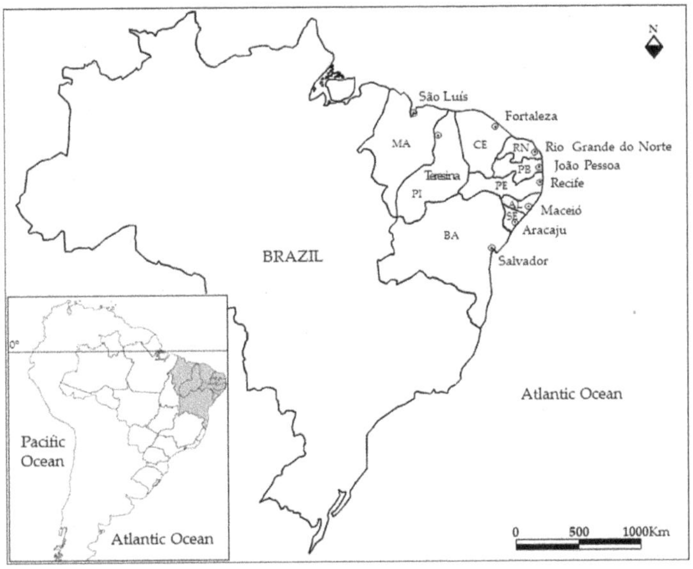

Figure 1. Map of the Brazil showing northeast region and sites described in the text.

2. Marine artisanal fisheries and fishermen in northeastern Brazil

2.1. Artisanal fishing

Fishing was an activity practiced by indigenous peoples prior to the arrival of the Portuguese. In the colonial period, European fishing practices were combined with native fishing practices, giving rise to different coastal cultures, such as the "*jangadeiros*" in the northeastern region, the "*caiçaras*" on the coast of the states of Rio de Janeiro and São Paulo in the southeastern region and the "*açorianos*" on the coast of the states of Santa Catarina and Rio Grande do Sul in the southern region [17].

In the 1960s, the Brazilian government implanted fishing industries mainly on the central-southern coast, which reached a peak in the 1970s, followed by a severe crisis the next decade, when the majority of these industries shut down [17]. In this period, industrial fisheries accounted for 64.5% of Brazilian fishery production [18]. However, this situation changed over the years. In 2006, artisanal fisheries accounted for 65.2% of the fishery production in the country and industrial fisheries accounted for 34.8% [19]. In northeastern Brazil, artisanal fisheries alone accounted for 96.3% of fishery production in the region in 2007 [12].

According to Brazilian Law n° 11959 issued on June 29th, 2009, artisanal fishing is defined as "any fishing activity practiced directly by a professional fisher in a self-employed fashion or in a familial economic regimen, using one's own means of production or landing through a partnership contract, which may involve the use of small watercraft" [20]. However, this definition does not consider the historic, social and cultural context of each community.

Artisanal fishing goes beyond a mere work activity and constitutes a way of life corresponding to social aspects that follow a behavioral pattern in traditional communities [21].

The artisanal fleet in the northeastern region is mainly comprised of sailboats (70.9%) due to the low production and maintenance costs as well as adaptation to the climate of the region, which has strong winds throughout nearly the entire year. Fishing operations are conducted in shallow waters near the coast and estuaries. However, motorboats account for the largest portion of fishery production (60.4%) due mainly to the greater autonomy, which allows longer fishing trips and consequently greater production [16].

Fishermen use a gamut of methods for exploiting the aquatic environment, as different species are targeted using a variety of technologies, which reflect local knowledge on the different environments in which these individuals live [22]. The gear is crafted artisanally using both local raw materials and industrial equipment, such as hooks and nylon nets [18, 23]. The type of fishing may be active (when the boat or gear is in motion) or passive (when the gear remains still throughout the duration of the practice) [16].

The main fishing gears used to catch fishes are gillnets, handlines and traps. Three types of gillnets are used: (1) a purse seine that encircles the school at the subsurface of the water; (2) a set gillnet, which is used to capture fish on the bottom or in the middle of the water column; and (3) a flying fish net, used to capture flying fish on the surface, which is only found in the state of Rio Grande do Norte. These fisheries occur in areas of shallow depths near the coast and involve trips of 0.5 to eight days. For fisheries targeting flying fish, however, distances of up to 35 miles and depths of approximately 1000 m have been recorded [16, 24].

Three types of handlines are used: (1) a bottom line with one to five hooks, weight and length adjustable to the depth, which can be as much as 250 m, with operations conducted 20 miles from the coast; (2) a surface line, on which secondary lines with hooks are attached to the main line above the fishing grounds; and (3) "corso" or "corrisco", which is similar to a surface line, but generally two lines (each with a one hook) are dragged as the boat moves across the water. This gear has the longest duration on the open ocean, with trips lasting 0.5 to 22 days and operations conducted 35 miles from the coast at depths as much as 1000 m [16, 24].

The use of a semi-fixed bottom trap has only been recorded in the state of Pernambuco. This type of trap has a hexagonal shape. It has a wood frame and is lined with mesh. The entrance is in the shape of a funnel. The fish enter the trap and cannot get out. This type of fishery is restricted to areas near the coast with a maximum depth of 80 m [16, 24].

Among all fishing gear, the handline accounts for the greatest fishery production due to the greater number and size of the catches. During the accompaniment of landings at 79 locations in seven states of northeastern Brazil between 1991 and 2001, handline operations accounted for 34.6% of the total production, followed by gillnets (27.3%) and trawl nets (10.8%) [16]. At landings in 16 locations in six states of the northeastern region between 1998 and 2000, 170 species from 52 families were identified, among which 149 species from 44 families were bony fishes and 21 species from eight families were cartilaginous fishes. The following are the main species targets in the different fisheries: *Seriola dumerili* (12.9%), *Mycteroperca bonaci* (11.6%) and *Lutjanus jocu* (7.4%) in bottom line operations; *Coryphaena hippurus* (10.3%), *Scomberomorus cavalla* (9.9%) and *Thunnus albacares* (3.1%) in surface line and "corrisco" operations; *Opisthonema oglinum* (23.1%), *Scomberomorus brasiliensis* (14.1%) and *Euthynnus alleteratus* (6.3%) in mid-water gillnet operations; *Carangoides bartholomaei* (5.6%), *Ocyurus chrysurus* (5.3%) and *Rhizoprionodon porosus* (5.3%) in bottom net operations; *Hemiramphus brasiliensis* (78.2%) and *Hyporhampus unifasciatus* (21.1%) in set gillnet operations; only *Hirundichthys affinis* in flying fish net operations; and *Pseudupeneus maculatus* in trap operations (45.6%) [24].

Corral fishing has also been recorded in the region, which involves a fixed trap made of stakes and wire similar to a fence, which imprisons fishes inside until their removal at low tide. These traps are deployed in regions where the continental shelf is wide and has a low incline. This practice is mainly found in the states of Ceará (91.1%) and Pernambuco (8.0%). In the state of Maranhão, only 10% of fishery production comes from corral fishing due to the high cost and need for constant maintenance. Fifty-seven species distributed among 26 families and 13 orders have been recorded, among which 22 species have commercial value in the region, such as *Scomberomorus brasiliensis, Centropomus parallelus, Cynoscion acoupa, Mugil curema* and *Megalops atlanticus* [16, 25].

The trawl net is bottom gear for catching shrimp that has structures, denominated otter boards, to maintain the net open. *Farfantepenaeus brasiliensis, Penaeus schmitti* and *Xiphopenaeus kroyeri* are the main species targeted in such operations [16]. The use of this gear is restricted to the mouths of important rivers due to the occurrence of reefs throughout nearly the entire coastal region [18], which impede trawling. The state of Alagoas accounted for the largest production of shrimp in the region between 1991 and 2001, followed by the states of Ceará and Sergipe [16].

Lobster fishing is performed over bottoms with calcareous algae on the continental shelf using three catch methods: (1) traps (locally denominated "covo" or "manzuá"), very similar to the fish traps employed in the state of Pernambuco, at depths of 10 to 40 m; this was the first method developed for catching lobster, but is currently little used due to its low degree of productivity; (2) dive fishing, in which the fisher dives with an air tank, catching lobsters in an active fashion; and (3) a gillnet made with multi-filament nylon 0.3 to 0.4 mm in diameter, buoys on the upper section and lead sinkers on the lower section at depths of 20 to 70 m [16]. Despite the fact that dive fishing and gillnet fishing are widely employed in the northeastern region, Brazilian law n 11.959 issued on June 29th, 2009 [20] prohibits these practices, which are considered predatory methods. In one community in the state of Rio Grande do Norte, 53.2% of the interviewees caught lobster using a gillnet, 20.6% performed dive fishing and only 9.5% deployed traps [21]. The same has been found in a community in the state of Ceará [26], demonstrating that illegal lobster fishing practices continue to be employed in the region due to the greater yields leading to overexploitation and unsustainability of this fishing.

The state of Ceará was the largest national lobster producer between 1991 and 2001, followed by the state of Rio Grande do Norte, with *Panulirus argus* and *Panulirus laevicauda* the main species targeted [16]. In recent decades, lobster stocks have suffered from overfishing due to the large number of vessels, illegal catch methods and catches of juveniles smaller than the permitted size [26]. Thus, this type of fishery has a defense period in the breeding season (December to May), when lobster fishing is prohibited by Law n 11959/2009 [20] and Normative Instructions n 138/2006 [27] and 206/2008 [28], with the fishing, transportation and sales of catches from illegal fishing practices considered an environmental crime (Law n 9605, Art. 34) [29].

The aforementioned characteristics of artisanal fisheries have been found in many locations of northeastern Brazil, such as one community in Pernambuco studied by the authors since 2006, one community in Rio Grande do Norte [30], two communities in Alagoas [31–32] as well as communities in the states of Bahia [21, 33] and Ceará [23, 26]. This demonstrates that, regardless of the location of the fishing community, the fishery characteristics in the northeastern region of the country are very similar, with no considerable operational differences. The variety of fishing gear and methods reflect the diversity of living resources found in different habitats as well as different types of bottoms and the currents that affect the region. This results in various multispecies fishing resources, which is the main characteristic of artisanal fisheries in the region.

2.2. Artisanal fishermen

Coastal areas in Brazil have very productive ecosystems, in which fishery and forest resources are important, as such environments ensure the survival of different human populations [34], including artisanal fishermen. Fishermen need to have empirical knowledge regarding natural environments, which was constructed through the process of appropriation of the environment during the practice of fishery activities both materially in the form of technologies and symbolically in the form of cognitive systems created in society [35]. Such knowledge involves understanding the environmental variables that affect fishing practices, such as tidal patterns, winds and seasonality, knowing how to manage fishing equipment as well as identifying the different forms of use and ecology of the targeted species, which constitutes vital information to the success of the activity. These characteristics have led to the denomination of artisanal fishermen as traditional peoples. According to Brazilian decree n° 6040 issued on February 7th, 2007 [36], "traditional peoples and communities are culturally differentiated groups that recognize themselves as such, have their own forms of social organization, occupy and utilize territories and natural resources as a condition for their cultural, social, religious, ancestral and economic reproduction, using knowledge, innovations and practices generated and transmitted through tradition." Communities such as artisanal fishing communities constitute an inestimable cultural heritage [37], but generally reside on the margins of society, overlooked by the government and public policies.

Marine artisanal fishing in Brazil is predominantly a male occupation [21–23, 38–43], which is mainly related to characteristics that favor men, such as the need for physical strength and long periods on the high sea. Women are culturally responsible for other tasks, such as household chores and raising children, which render marine fishing impossible. When women participate in artisanal fishing, they are more involved in shellfish gathering, which is an important activity in the estuaries of northeastern Brazil.

Marine artisanal fishing involves individuals of all ages, from youths to the elderly. In the city of Recife (state of Pernambuco), 80.5% of fishermen are adults [43]. In the community of Itacaré in the state of Bahia, 68% of fishermen are between 26 and 45 years of age [33]. In a rural community in the state of Rio Grande do Norte, 54% of fishermen are between 31 and 50 years of age [21]. The striking presence of adults in fishing communities in the economically active age range confirms the importance of fishery activities to the local economy.

Fishermen generally have a low degree of schooling [43–46]. However, advances in schooling between generations have been found in a community in the city of Recife, where most fishermen have incomplete elementary (26.1%) or high school (25.0%) educations, 61.1% of their parents are illiterate and 25.4% of their descendants have a complete high school education [43]. However, this is not the case in the state of Bahia, where 40.0% of fishermen remain illiterate [44], or the state of Rio Grande do Norte, where 78.8% have an incomplete elementary or middle school education and only 4.7% have a complete high school education [21]. With the reduction in the illiteracy rate in Brazil [47], some urban fishing communities have a higher level of education, which may be related to the proximity to large cities, where schools are more accessible. The low degree of schooling among fishermen exerts a direct effect on the social organization of this class of laborers [42], which may result in the unawareness of workers' rights, thereby weakening this category of professionals and causing both disadvantages and negative repercussions regarding the negotiation of the price of the catch.

Artisanal fishermen generally have ties with the community in which they live and a longer period of time spent in such communities leads to a greater feeling of belonging [48]. This characteristic is fundamental to strengthening the activity, as a united community faces difficulties with greater ease. In the state of Bahia, fishermen reside an average of 40 years in the rural community [44]. In a community in the city of Recife, a large portion of fishermen have resided in the location since birth (42.0%) or since the creation of the neighborhood (10.2%) [43], which demonstrates that, even in an urban community, the influence of urbanization has not yet changed this important aspect of artisanal fishing.

The type of residence varies considerably with the socioeconomic status of each community. For fishermen in an urban area of the city of Recife, most homes are made with bricks and mortar [43, 45]. In a rural community in the state of Paraíba, a large portion of residences are made of wood and clay [48], but this form of home is being replaced by the practicality and durability of bricks and mortar in the capital of the state [49]. In a rural community in the state of Rio Grande do Norte, the majority of homes were made with sticks and mud. However, the municipal authorities benefitted the residents by constructing homes made of bricks and mortar, thereby changing the face of the community [30].

Sanitation is another factor that varies with the characteristics of the location. Rural fishing communities are generally isolated, with inadequate living conditions, such as the coastline or areas near mangroves, where sanitation conditions are often precarious [46, 48]. For the urban fishermen in the city of Recife [43] and the state of Paraíba [49], the conditions might be considered better. However, a similar situation is reported. Even with more frequent trash collection, most urban waste is dumped into rivers and estuaries near fishing communities, which does not differ from many large cities in Brazil with a sewage system, the treatment of which is incomplete and precarious [50], demonstrating that the deficiency in sanitary conditions is not only found within fishing communities.

Fisheries are important to coastal communities due to the large number of direct and indirect occupations created in the different sectors of the productive chain [37]. In one community in the state of Pernambuco, mean monthly income was higher than the minimum wage at the time, with higher salaries for those with their own fishing vessels. However, 20.9% exerted other activities to complement their income [43]. In communities in the state of Bahia, 55% of fishermen exercise other activities to increase their income [33]. The same is reported for communities in the state of Ceará [21, 26]. Despite the economic importance of fishery activities, income varies based on the yield and many fishermen seek other activities to complement their income, which demonstrates the low yields and economic under-valuation of artisanal fishing practices in northeastern Brazil.

3. Transmission of traditional knowledge and urban pressure

Artisanal fishermen reveal complex knowledge acquired through tradition inherited from older fishermen and contribute to the maintenance and sustainable use of natural ecosystems [51]. Such knowledge is generally transmitted orally from father to son and carefully guarded by fishermen [51] as a precious inheritance to ensure family survival and the continuation of the activity. The family is an important source of the transmission and maintenance of traditional knowledge and practices. Indeed, family ties constitute the main transmission route of knowledge regarding fishing in artisanal communities. The empirical knowledge of fishermen involves information of considerable cultural and ecological value that is essential to the continuation of local customs and the investigation of natural resources. This traditional knowledge can be used as a source of data for the establishment of species management and conservation plans, as it generally corresponds to scientific knowledge. In some countries, social ecology has used the knowledge of traditional communities for conservation purposes, as the strong dependence on natural resources, symbolic structure, management systems developed over long periods of time and often isolation make traditional fishing communities partners in conservation efforts [51].

The knowledge of fishermen involves navigation and the identification of fishing grounds through triangulation systems as well as the diversity, seasonality, migration habits and feeding habits of fish, fishing methods, depths and types of ocean bottoms [51]. Knowledge on the feeding habits of fish is evident in the community of Brasília Teimosa in the city of Recife (state of Pernambuco), where fishermen know which type of gear to use to enhance the success of fishery production [52].

In the community of Itapissuma near the coast in the state of Pernambuco, fishermen have precise knowledge of the estuary-mangrove complex, which allows social production and reproduction, maintaining a cultural heritage in which fishing practices are transmitted orally between generations and through both observations and practical learning [53]. In the rural community of Itacaré in the state of Bahia, most fishermen learned their craft in childhood, especially from fathers and family members. This demonstrates that knowledge regarding fishing practices and the use of natural resources can still be transmitted from generation to generation, thereby maintaining the characteristics of artisanal fishing [33].

Although rooted in tradition, artisanal fishermen have also been transformed by internal and external dynamics, but at a slower pace than that found in urban societies [51]. Traditional communities customarily maintain local knowledge accumulated and constructed over the years, which allows close contact with nature and the use of natural resources, even in communities located in urban areas, where the environment, albeit changed, allows sustaining modes of living based on the use of natural resources. However, this knowledge undergoes constant pressure from the urban way of life and can become lost over time due to the characteristic of the oral transmission as well as the contrast between traditional management practices and technologies available in the urban environment [54].

The urban community of Poti Velho in the city of Teresina (state of Piauí) continues to preserve its own culture, in which traditional knowledge is passed on from generation to generation, transferred within the family, but undergoes constant pressure from the urban way of life [54]. The community of Vila Velha in the state of Pernambuco entered into a stage of the transformation of habits in 1999 due to the increase in tourism and real estate speculation. Knowledge related to fishing was passed on from generation to generation by family members and more experienced fishermen. However, the continuity of these practices is rather doubtful, as the increase in the level of schooling of the descendants offers different professional opportunities to future generations [55].

The community of Brasília Teimosa, which is located in an urban area in the city of Recife (state of Pernambuco), has been the object of study of the authors of the present investigation for several years and changes have been observed in this community. In studies conducted since 2006, the family constituted an important source for the transmission and maintenance of knowledge and traditional practices, as most fishermen report having learned this knowledge from fathers, uncles and brothers, with the constant presence of relatives in the activity. However, a small number of the descendants of fishermen become involved in fishing practices, which suggests that the descendants may not give continuity to the fishing tradition. This situation is aggravated, as the relationship between father and son was the most important form of the transmission of knowledge regarding fishing practices in Brasília Teimosa [52]. Thus, the community may be losing the traditional form of knowledge transmission within the family, which can change the face of its traditional nature. As this community is situated in an urban area in the city of Recife, residence in Brasília Teimosa allows a set of job opportunities in other sectors, which may be more attractive to young people due to the advantages offered as well as social benefits. As a result, artisanal fishing is currently being threatened as an economic and social activity in this area.

Porto de Galinhas Beach in the city of Ipojuca is the most visited touristic point in the state of Pernambuco and was a community decades ago in which artisanal fishing was an important activity. This community has been undergoing constant change due to the economic growth stemming from tourism. The authors of a study conducted in the community found that most fishermen had descendants who worked in local stores or in the tourism industry, whereas only 12% practiced fishing with local traditional gear [56]. Besides the disorderly human occupation on the coast due to tourism, the destruction of mangroves and riparian forests due to real estate speculation and agricultural activities has also contributed to a reduction in fish stocks [56], further leading to the avoidance of fishery activities on the part of youths. Thus, there is no renewal of the group with the inclusion the descendants of fishermen, which constitutes a barrier to the transmission of traditional knowledge, leading to the risk of this community not surviving another generation [56].

In urban fishing communities in the state of Paraíba, the disbelief among fishermen regarding improvements in working conditions and quality of life has led these traditional laborers to not want their descendants to enter the fishing profession, explaining that it would be better for them to study and become qualified for professions available in the urban setting in which they live [57]. This demonstrates a degree of disinterest among these fishermen in passing on knowledge of the activity to the next generation. In such urban communities, with the diminished fish stocks in recent decades, fishing is no longer advantageous in comparison to other urban opportunities and has become a secondary activity [57] that does not attract the interest of younger generations.

In São Francisco do Conde, which is an urban area in the city of Salvador (state of Bahia), studies report that the descendants of fishermen prefer to study and learn a different profession, with the encouragement of their parents and government assistance in the form of scholarships, thereby becoming unaware of fishing knowledge and leaving this activity as an option for those who have no other opportunities [58]. In Fernão Velho, which is an urban community in the city of Maceió (state of Alagoas), 30% of interviewees learned to fish from family members, but do not want their descendants to be fishermen, which demonstrates that cultural components are no longer being transmitted to younger generations. The intensive changes in the natural landscape due mainly to urbanization and industrialization have led to a reduction in native vegetation and fish stocks, with a direct reflection on socioeconomic aspects. While many traditional fishermen and their descendants no longer consider fishing an attractive profession, unemployed individuals enter into this activity, which indicates a likely marginalization of fishery practices and the risk of the extinction of the artisanal nature of this community [59].

In the municipality of Cabo de Santo Agostinho (state of Pernambuco), the Suape region has also been undergoing changes due to industrialization and tourism. This region has been the object of studies conducted by the authors of the present investigation since 1997. Suape Bay was a native coastal region in past decades and considered one of the most important marine and estuarine areas on the coast of northeastern Brazil. Moreover, one of the main artisanal communities in the state is situated in this area. With the implantation of the Suape Port and Industrial Complex beginning in 1979 and the subsequent creation of a large resort on the bay, changes have occurred in local geomorphology and hydrodynamics. In studies conducted in 1997 and 1998, local fishermen reported that the implantation of these two enterprises was causing a reduction in local fishery production, thereby compromising the income of fishermen and leading to the avoidance of the profession on the part of younger generations. This fishing community is currently much smaller, as most descendants of fishermen are not interested in the activity and have entered other professions, indicating the likely extinction of this traditional community in upcoming generations.

The urbanization process may not initially affect the traditional nature of fishing communities, but exerts an influence on ways of thinking as well as the very characteristics of fishery activities, compromising the maintenance and continuity of such activities by impeding the transmission of knowledge. As the perpetuation of fishery knowledge occurs orally, with no records to ensure the practice of operations, catch techniques, knowledge regarding fishing grounds and fish species, the maintenance and sustainability of marine artisanal fisheries in urban areas as well as the communities themselves are seriously affected due to the pressures imposed by the new developmentalism.

4. Territoriality and urban pressures

The immense socio-cultural diversity of Brazil has been accompanied by an extraordinary land diversity that includes different forms of territoriality maintained by traditional communities and populations [60]. Thus, the term "territoriality" is defined as a collective effort of a social group to occupy, use, control and identify itself with a specific portion of its biophysical environment, converting it into the group's "territory" or homeland [61]. However, a three-dimensional concept of space is found in the cultural tradition of artisanal fishermen, which encompasses distinct domains of life (ocean, land and sky) imbued with meaning [62], giving such populations a broader notion of territory.

Although artisanal fisheries involve extractist activities conducted in a family economy regimen with specific means of production and striking characteristics in the profile of the laborers, urban growth has been affecting these traditional communities and transforming their forms and functions, which requires new models for understanding the changes imposed by urbanization [57], such as real estate speculation, a reduction in living and working space, disorderly tourism practices, pollution and social exclusion. Rapid urbanization along coastal areas also results in the emission of effluents. The implementation of industrial centers has led to a breakdown of the natural productivity of ecosystems and fishing practices [63], with consequent environmental degradation and the loss of quality of life in traditional communities [64].

The tendency toward occupying coastal areas in Brazil has substantially aggravated the impact on artisanal fishery activities. Such areas hold a strategic position in commercial exchanges, concentrating port activities, favoring the establishment of cities and industries and offering numerous recreational attractions to large urban masses [65]. This process has led to the shrinking of fishing areas and territories, with the loss of launching and landing locations as well as fishing grounds.

Northeastern Brazil accounts for 18% of the national territory and has nine coastal states, the most evident characteristic of which is the large concentration of the population along the coast, resulting from centuries-long occupation related to commercial relations with other countries [66–67]. Prior to Brazil's incorporation into the process of globalization, native populations disputed their territorial spaces with real estate developers. Fishermen were forced to form small communities with no basic infrastructure or urban services or began to reside in the periphery of nearby urban centers, also without infrastructure or services [67]. One such example is the real estate pressure that occurred in the 1970s on the urban beaches of Tambaú and Lucena in the city of João Pessoa (state of Paraíba) and currently occurs on Penha Beach in the same city. This is due to the peripheralization process of richer classes of society, who have exchanged residences in the center of town, which has since become more directed toward commerce, for spaces occupied by fishing communities, which are currently highly prized by the real estate industry, thereby compromising the ability of fishing communities to remain in coastal areas [57]. In the neighborhood of Brasília Teimosa in the city of Recife (state of Pernambuco), where the main office of the Z-01 Fishers' Colony is found and most artisanal fishermen of the city reside, there are longstanding reports of pressure to remove the fishing community from the area. This struggle for the right to use the land, defined as "stubbornness", coincided with the construction of the federal capital (Brasilia) in the center of the country and gave rise to the name of the community: Brasília Teimosa (Stubborn Brasilia) [68–69]. This community continues to face real estate pressure and other social pressures stemming from the economic growth of the state of Pernambuco [43].

Tourism and real estate speculation also exert significant pressures on fishermen in rural areas of northeastern Brazil. Rural communities, such as Prainha do Canto Verde on the coast of the state of Ceará, exhibit much greater strength and organization. Thirty years ago, a Brazilian Court granted this community the recognition of its right to land against a real estate company that declared itself the proprietor of the land. To remain, the inhabitants combined fishing with community tourism managed by the social group itself, which established operational rules, such as the prohibition of sales of land to individuals who were not part of the community [67]. This demonstrates community ties with the location and a feeling of belonging.

Besides the pressure regarding territory due to real estate speculation and tourism, the expansion of capital due to enterprises such as shrimp farming also results in a change in the land structure of rural fishing communities. In Canto do Mangue in the state of Rio Grande do Norte, lands located on the banks of the estuary, which were previously leased to fishermen for family agricultural purposes, began to be the target of private initiatives for the shrimp farming industry with the raising of the exotic species *Litopenaeus vanammei*, which besides occupying areas used by artisanal fishermen, has contributed to the reduction in fish stocks as well as the accelerated degradation of mangroves [30]. This type of activity also leads to the discharge of toxic effluents in bodies of water, the expulsion of the families of fishermen from their places of residence and the restriction of their access to traditional fishing grounds [63], thereby imposing the loss of territories.

Artisanal fisheries are also affected by offshore oil exploration, which brings serious risks to the coast of the states of Sergipe and Bahia, the severe pollution from alcohol production plants, which is released directly into estuaries, especially in the states of Pernambuco and Alagoas, and the discharge of toxic waste into lagoons and bays of high natural productivity, such as Manguaba lagoon in the rural area of Marechal Deodoro (state of Alagoas), Mandaú lagoon in the urban area of the city of Maceió (state of Alagoas) and Todos os Santos Bay in the city of Salvador (state of Bahia) [63]. Todos os Santos Bay is an environmental protection area surrounded by a metropolitan area with port and industrial activities, including an oil refinery

[70], where an important fishing community is being negatively affected by the new developmentalism.

Even with the social changes occurring in fishing communities in recent decades due to the expansion of urbanization, tourism and real estate speculation, artisanal fishermen constitute social subjects that have a form of spatiotemporal ordination that is dissonant with the urban-industrial context and have a heritage stemming from their centuries-long interaction with nature, which modernity cannot disregard [62]. Through material and immaterial cultural manifestations of fisheries, the activity may still last, as many individuals, especially tourists, enthusiasts of nautical trips and amateur fishers may pay to experience the work and way of life of traditional fishermen. Moreover, the traditional community in the urban setting may also be considered an obligatory part of social reports for environmental licenses granted to large enterprises [57], especially those that affect fishing territories.

The environmental impact report (denominated RIMA) by Brazilian environmental agencies for enterprises of the Marine Waterfront Recovery Project in the cities of Jaboatão dos Guararapes, Recife, Olinda and Paulista (state of Pernambuco) is an example of the participation of urban fishing communities in reports for environmental licenses. This report involved the participation of 50 leaders who answered questionnaires and 203 fishermen who participated in meetings and workshops to report on the impact of enterprises [71]. The fishermen were only considered in the final phase of the project, but should have been consulted during its design, as the numerous negative impacts are generally not mitigatable for artisanal fisheries.

The meaning of the surrounding environment is fundamental to artisanal fisheries, as fishermen maintain interactions with natural resources and their territory, even if the activity is threatened by different forms of pressure as well as environmental and social changes. Thus, understanding sustainability in this situation requires a new look at social practices, in which fishers should have greater participation, as these men and women struggle to maintain their identities, ways of life, territories and social visibility.

5. Public policies and the sustainability of artisanal fisheries

5.1. Public policies directed at artisanal fishing in Brazil

Historically, Brazilian public fishery policies were designed with no consideration for traditional communities and were based on the modernization of the exploitation of natural resources [72], with a position clearly in favor industrial fishing activities and large enterprises based on the dualism of the old *versus* the new, with small-scale fishermen considered reactionary, uncultured, predatory and incapable of assimilating new technological standards [73]. In the minds of the proponents of modernity, the tradition of professional artisanal fishing as work and a way of life no longer performs any efficient role in human development and is even seen as the cause of different forms of environmental degradation due to the fact that it is an activity the uses natural resources. Such an equivocal argument could be clarified if these proponents of modernity accompanied fishery routines [74].

In Brazil, fishery policies after World War II followed a worldwide tendency of growth based on industrialization that dominated the 1950s and 60s. The Superintendence of Fishery Development [Superintendência do Desenvolvimento da Pesca (SUDEPE)] was created in October 1962 for the regulation of fisheries and fishing resources. During the time when fisheries were regulated by SUDEPE, the priority was the modernization and industrialization of fishing activities and catches went from 300,000 tons in the 1960s to 900,000 tons between 1970 and 1980 [75]. However, this massive incentive for industrial fishing practices led to the decline of stocks and the failure of businesses linked to the fishery sector [75].

With the failure of the adopted public policy, SUDEPE was extinguished in 1989, passing the control of fisheries and fishing resources to the Brazilian Institute of the Environment and Renewable Resources [Instituto Brasileiro do Meio Ambiente e dos Recursos Renováveis (IBAMA)]. In 1998, the Fishery and Aquiculture Department was created (Decree n° 2.681 of June 21st), linked to the Ministry of Agriculture and Water Supply, and, together with IBAMA, served as the regulating agency of fishery activities.

To optimize the regulation of fishery operations, the Special Secretary of Aquiculture and Fisheries was formed by the Federal Presidency through Provisory Measure n° 103 (January 1st, 2003) and later became the Ministry of Fisheries and Aquiculture (MPA) through Law n° 11.958 (June 26th, 2009) issued symbolically on June 29th, 2009, which is Fishers' Day in Brazil [76]. The aim of the MPA was to "foster and develop policies directed at the fishery sector in conjunction with its aspirations. [...] founded in the marks of a new management policy, maintaining a commitment with environmental sustainability regarding the use of fishery resources" [76].

Due to the increase in domestic and industrial pollution, real estate speculation and the few possibilities of a significant increase in fishery production, except for marine fisheries, the Brazilian government proposed a credit policy for the fishery sector in 2003 to encourage the conversion of artisanal fishing to aquiculture, arguing that this would be an alternative for fishermen due to the overexploitation of wild fishes [74]. However, there was no participation on the part of artisanal fishermen in the stock evaluation process and no consideration of regarding their traditional knowledge. Thus, wishing to modernize the fishery sector, the Brazilian government ended up promoting the breakdown of fishing community traditions [74], mainly by not adopting a participative management plan that considered traditional knowledge accumulated and passed down from generation to generation.

During the existence of the SEAP and MPA, fishery management made some advances in public policies directed at the profession, with the revising or implantation of laws to ensure benefits and rights for fishers, such as (1) economic subsidies for diesel oil used for fishing vessels (Law n° 9445, March 14th, 1997) [77]; (2) the General Fishery Register related to licenses, permission and authorization granted for fisheries and aquiculture (Law n° 10683, May 28th, 2003; Normative Instruction n° 6, May 19th, 2011) [78–79]; (3) unemployment insurance during the fishery defense period (Law n° 10.779, November 25th 2003) [80]; (4) rural credit for family enterprises among small-scale farmers and artisanal fishers (Law n° 11326, July 24th, 2006) [81] through the National Family Agriculture Strengthening Program; (5) the recognition of fishery colonies as well as state and national fishers' federations as agencies of the class of laborers of artisanal fisheries (Law n° 11699, June 13th, 2009) [82]; and (6) the New Fishery Law (Law n° 11959, June 29th, 2009) [20], which lays down the National Aquiculture and Fishery Sustainable Development Policy. Besides these policies, social programs of the government, such as the Family Grant, have assisted in improving the living conditions of a portion of fishermen [83]. However, during fishery management by the SEAP and MPA, many aspirations of fishermen were not considered, as artisanal fishermen are often impeded from using government benefits due to not meeting the required criteria or for not being part of the official Brazilian statistics regarding fisheries. As a result of current political affairs in Brazil, the MPA was extinguished on October 2nd, 2015 (Provisory Measure n° 696/2015) [84], when fisheries and aquiculture began to be regulated by a secretary linked to the Ministry of Agriculture, Livestock and Water Supply. The discontinuity of the MPA has generated uncertainties regarding the new direction of fisheries.

5.2. Sustainability of artisanal fisheries

Artisanal fisheries account for more than 90% of fishing jobs and the catches of this modality represent more than half of fishing catches throughout the world [85]. A worldwide decline in fish stocks [86–88]) as a result of unsustainable practices and an increase in fishing efforts [89–90] have been well documented in recent decades. Moreover, the global marine fisheries catches reported by FAO are underreported, being captured 30% more than is declared, with a peak catch of 130 million tons in 1996 and has been declining more strongly since [91].

In Brazil, artisanal fishing also faces a general lack of biological, socioeconomic, technological and organizational information, resulting from the dispersion and complexity of the activity. The difficulty is even greater when considering the variety of multispecies gear and the diversity of the resources captured. The insufficient information is evidenced by the lack of political attention directed at the profession, which is often a reflection of the conventional approach with a focus on biological aspects and no consideration given to the economic and institutional aspects of fishing communities, which contributes the invisibility of artisanal fisheries with regard to public policies [18] as well as the lack of sustainability of this modality.

Considering the worldwide decline in catches, efforts have been made to implement global actions directed at the sustainability of fisheries based on (1) the Code of Conduct for Responsible Fisheries [92] established in 1995 by the Food and Agriculture Organization (FAO) of the United Nations aimed at the ecological and social sustainability of fisheries; (2) the principles of the ecosystem approach to fisheries [93] aimed at planning, developing and generating fisheries to meet the different needs and desires of societies, benefiting from the complete variety of goods and services provided by marine ecosystems; and (3) the RAPFISH method, which is a multidisciplinary rapid appraisal technique for evaluating the comparative sustainability of fisheries [94–95]. In Brazil, the RAPFISH method has been used in eight states from north to south. When fisheries were compared jointly, sustainability indicators did not reveal any clear patterns, demonstrating that similar approaches will be needed in the future to assist fisheries in Brazil [96], and RAPFISH is a method that undergoes a constant improvement process.

Other global actions with an impact on sustainable fishing are (1) labeling and certifications for sustainable fishing products, such as the eco-labeling guidelines for fish and fishery products from marine fisheries designed by FAO [97] to certify and promote labels for products from well-managed marine fisheries and focus on issues related to the sustainable use of fishery resources; (2) the certificate from the International Social and Environmental Accreditation and Labelling Alliance (ISEAL Alliance), founded in 2002, which is an association of leaders of international organizations that establishes standards for social and environmental issues and certifies products, including fish catches, that comply with the ISEAL good practices code [98], involving open, transparent, participative processes with a proven standard of credibility and measures to ensure that even the most marginalized interested parties have something to say about the development of standards; and (3) the certification of the Marine Stewardship Council (MSC), which is an international non-profit organization aimed at contributing to the health of oceans that certifies fishery based on sustainable practices [99]; the MSC is a pioneering agency that maintains dialogs with all fishery sectors as well as a broad spectrum of stakeholders [100]. Sustainable fishing is the aim of such labeling and the MSC certification has had a positive impact on the environment, but has marginalized fisheries, especially those in low-income countries, as the council focuses on the sustainability of fish stocks and not fisheries [101].

Besides these important guidelines, fishery management must also consider the actions of the United Nations Millennium Development Goals, the aims of which are the eradication of poverty on the global scale, problems such as bycatch, habitat loss, species introductions and invasive species as well as the globalization of the fishery market [102]. Regardless of global guidelines, each nation adopts its own public policies and presents specific regulations based on its particular situation. In Brazil, the bases of fishery sustainability are founded on the National Policy for the Sustainable Development of Fishery Activities (Law no. 11959/2009) [20]. To achieve a balance between the principle of the sustainability of fishery resources and the obtainment of both economic and social results, this policy establishes access regimens, total permissible catches, sustainable fishing efforts, defense periods, fishing seasons, catch sizes, protected areas, reserves, gear, methods, fishery systems, aquiculture systems, the support capacity of the environment, fishery monitoring/control and the protection of fishes in the process of reproduction or the re-composition of stocks. Section 1 of this law declares that fishery management should consider the peculiarities and needs of artisanal fishermen, subsistence activities and family farming to ensure the continuity of these aspects and practices. In 1995, the Code of Conduct for Responsible Fisheries [92] officially recognized the need to consider the traditional knowledge of artisanal fishermen regarding fishery resources.

With the growing crisis in recent years [86, 103] and the limited capacity to predict complex systems, such as the marine ecosystem, together with the fact that fishery management should be conducted in a broad-scoped, integrated fashion to maintain the productive capacity and resilience of linked social/ecological systems [102, 104], there has been a recent increase in information, indicating the importance of including traditional knowledge in fishery management. However, despite the fact that the importance of such knowledge is frequently expressed, it has not been observed in practice in Brazil, as demonstrated in a recent study through the reports of artisanal fishermen in the communities of Baldo do Rio and Carne de Vaca (state of Pernambuco), who state that political actions are distant from their interests [105]. Urban artisanal fishermen, in the state of Pernambuco, are also affected by the lack of attention given by institutions linked to fisheries [106].

The implantation of marine protected areas and marine extractive reserves is recognized as an important tool for fishery sustainability. The aim of such areas is to achieve the conservation and sustainability of fisheries, thereby contributing to biodiversity and habitat conservation as well as other ecological/social benefits beyond the boundaries of these areas [107]. In Brazil, such areas are considered a community-based, site-specific, multi-use, land and sea resource management approach based on claims of culturally distinct groups with longstanding livelihood ties to 'artisan-scale' production territories [108]. Thus, it is essential to consider social issues and long-term benefits for the effective management of these areas [109]. Studies conducted in Brazil indicate that such protected areas have been established without consulting artisanal fishermen and that fishery management plans are designed by scientists without incorporating the traditional knowledge of fishermen, thereby increasing social marginalization and the loss of cultural identity, whereas management involving fishermen is crucial to the success of the project [109]. Therefore, the success of this tool in Brazil remains dependent on managers willing to include all stakeholders in the implementation of protected areas.

Sustainable fishing remains a goal to be reached that should involve inclusive governance, encompassing the complexity of the ecosystem and its natural uncertainties, such as climate change, the dynamics of the market in the light of growing globalization and all stakeholders involved in fisheries, with particular inclusion of the protagonists of this activity – artisanal fishers.

6. Conclusion

Artisanal fishing is a longstanding activity responsible for numerous direct and indirect jobs and the largest portion of fishery production in northeastern Brazil. The modality is characterized by a variety of gear and techniques that reflect the diversity of living resources found in different habitats, resulting in fishery activities with multispecies resources. To exploit such resources, artisanal fishermen rely on rich traditional knowledge regarding the environment and fishery resources, which is fundamental to the maintenance of the way of life of these communities and is directly responsible for the success of fishery practices. Despite the economic importance of fisheries, many fishermen seek other activities to complement their income. This underscores the low yields and economic devaluation of artisanal fishing in northeastern Brazil, which, together with the pressure of the new developmentalism in urban areas, have been causing changes in traditional fishing communities. Pressure from real estate speculation, the reduction in fishing territories, disorderly tourism and pollution have led to a drop in the natural productivity of coastal ecosystems and fishery production, thereby compromising the quality of life and maintenance of urban fishing communities.

Moreover, the process of urbanization affects ways of seeing and thinking as well as the very characteristics of fishery activities, such as the transmission of knowledge, since younger generations are not interested in fishing and look to more attractive professions offered by large urban centers. As the perpetuation of knowledge occurs through oral practices, with no written records to ensure continuity, the maintenance and sustainability of marine artisanal fishing in urban areas as well as the communities themselves are seriously compromised in the face of the pressures imposed by the new developmentalism. Besides urban pressures and declines in fish stocks, which contribute to the discontinuity of urban fishing communities, the Brazilian federal government does not strengthen traditional fishing communities and adopts actions that fail to take traditional knowledge into consideration, thereby contributing to the breakdown of the transmission of the activity as well as the social invisibility of these laborers.

Sustainable fishing remains a goal to be reached that should involve inclusive, participatory governance, encompassing the complexity of coastal ecosystems, the sustainability of fisheries and fishery resources as well as the maintenance and continuity of traditional communities.

Author details

Simone F. Teixeira[1,2*], Daniele Mariz[2,3], Anna Carla F. F. de Souza[2,3] and Susmara S. Campos[2,4]

*Address all correspondence to: teixeirasf.upe@gmail.com

1 University of Pernambuco, Recife, Pernambuco, Brazil

2 Research Group on Ethnoecology and Tropical Fish Ecology Studies of University of Pernambuco, Recife, Pernambuco, Brazil

3 Federal University of Pernambuco, Recife, Pernambuco, Brazil

4 Federal Institute of Education, Science and Technology of Pernambuco, Recife, Pernambuco, Brazil

References

[1] Brasil. Constituição da República Federativa do Brasil, de 5 de outubro de 1988. [Internet]. 1988. Available from: http://www.planalto.gov.br/ccivil_03/Constituicao/Constituicao.htm. [Accessed: 2016-01-04]

[2] Brasil. Decreto n 5.300, de 7 de dezembro de 2004 [Internet]. 2004. Available from: http://www.planalto.gov.br/ccivil_03/_ato2004-2006/2004/decreto/D5300.htm [Accessed: 2016-01-04]

[3] Brasil. Lei n 8617, de 4 de janeiro de 1993. [Internet]. 1993. Available from: http://www.planalto.gov.br/ccivil_03/LEIS/L8617.htm [Accessed: 2016-01-04]

[4] Marinha do Brasil. Amazônia azul [Internet]. 2004. Available from: https://www.marinha.mil.br/sic/amazonia-azul.html [Accessed: 2016-01-04]

[5] Coutinho PN. Levantamento do estado da arte da pesquisa dos recursos vivos marinhos do Brasil. Relatório do Programa REVIZEE (1995–2000), Oceanografia Geológica. Brasília: FEMAR/SECIRM/MMA; 2005. 138 p.

[6] Freire GSS, Manso VAV, Silva Filho, WF, Gomes, DF, Lima, SF, Bezerra, SRP, et al. Sedimentologia e Morfologia da Margem Continental do Nordeste do Brasil . In: Hazin FHV, editor. Meteorologia e Sensoriamento Remoto, Oceanografia Fisícal, Oceanografia Química e Oceanografia Geológica (v. 1). Fortaleza: Martins & Cordeiro; 2009. pp. 217–45.

[7] Pickard GL, Emery WJ. Descriptive physical oceanography: an introduction. 5th ed. Oxford: Elsevier; 1990.

[8] Nonaka RH, Matsuura Y, Suzuki K. Seasonal variation in larval fish assemblages in relation to oceanographic conditions in the Abrolhos Bank region off eastern Brazil. Fish Bulletin. 2000;98:767–84.

[9] Ekau W, Knoppers B. An introduction to the pelagic system of the North-East and East Brasilian shelf. Arch Fish Mar Res. 1999;47(2/3):113–32.

[10] Travassos P, Hazin FHV, Zagaglia JR, Advincula R, Schober J. Thermohaline structure around seamounts and islands of northeast Brazil. Arch Fish Mar Res. 1999;47(2/3):211–22.

[11] Lessa R, Bezerra Jr. JL, Nascimento ED, Lima M, Pereira AA. Oceanografia biológica: composição, distribuição e abundância do ictioneuston na ZEE da região Nordeste do Brasil. In: Hazin FHV, editor. Biomassa fitoplanctônica: Biomassa primária e secundária, macrozooplâncton, ictioplancton, ictioneuston, macrofauna bêntica (v. 2). Fortaleza: Martins & Cordeiro; 2009. p. 166–94.

[12] Castello J. O futuro da pesca da aquicultura marinha no Brasil: a pesca costeira. Cienc Cult. 2010;62(3):32–5.

[13] Ferreira BP, Maida M, Cava F. Características e perspectivas para o manejo da pesca na APA marinha Costa dos Corais. In: Congresso Brasileiro de Unidades de Conservação: Anais do II Congresso Brasileiro de Unidades de Conservação; 2000 Nov 5–9; Campo Grande, Brasil. Curitiba: Rede Nacional Pró-Unidades de Conservação. 2000. p. 50–8.

[14] MMA (Ministério do Meio Ambiente - Gerência de Biodiversidade Aquática e Recursos Pesqueiros). Panorama da conservação dos ecossistemas costeiros e marinhos no Brasil. Brasília: MMA/SBF/GBA; 2010. 148 p.

[15] Scherer M, Sanches M, Negreiros DH. Gestão das zonas costeiras e as políticas públicas no Brasil: um diagnóstico. In: Barragán Muñoz JM, coord. Manejo costero integrado y política pública en Iberoamérica: Un diagnóstico. Necesidad de cambio. Cádiz: Red IBERMAR (CYTED); 2010. p. 291–330.

[16] Lessa R, Bezerra Jr. JL, Nóbrega MF. Dinâmica das frotas pesqueiras da região nordeste do Brasil (v. 4). Fortaleza: Editora Martins & Cordeiro; 2009. 164 p.

[17] Diegues AC. A sócio-antropologia das comunidades de pescadores marítimos no Brasil. Etnográfica. 1999;3(2):361–75.

[18] Vasconcellos M, Diegues AC, Sales RR. Limites e possibilidades na gestão da pesca artesanal costeira. In: Costa AL, editor. Nas redes da pesca artesanal. Brasília: IBAMA; 2007. p. 15–63.

[19] IBAMA. Estatística da Pesca 2006 Brasil: grandes regiões e unidades da federação. [Internet]. 2008. Available from: http://www.ibama.gov.br/documentos-recursos-pesqueiros/estatistica-pesqueira. [Accessed: 2016-01-12].

[20] Brasil. Lei n° 11959, de 29 de junho de 2009. [Internet]. 2009. Available from: http://www.planalto.gov.br/ccivil_03/_ato2007-2010/2009/Lei/L11959.htm. [Accessed: 2016-01-18].

[21] Castro FDD. Implicações socioeconômicas e ambientais da pesca artesanal de lagosta em Touros/RN. [dissertation]. Natal: Federal University of Rio Grande do Norte; 2013.

[22] Souza MF. Etnoconhecimento caiçara e uso dos recursos pesqueiros por pescadores artesanais e esportivos no Vale do Ribeira. [dissertation]. Piracicaba: Federal University of São Paulo; 2004.

[23] Costa-Neto EM, Marques JGW. Conhecimento ictiológico tradicional e a distribuição temporal e espacial de recursos pesqueiros pelos pescadores de Conde, Estado da Bahia, Brasil. Etnoecológica. 2000;4(6):56–68.

[24] Nobrega MF, Lessa RP. Descrição e composição das capturas da frota pesqueira artesanal da região nordeste do Brasil. Arq Ciências do Mar. 2007;40(2):64–74.

[25] Piorski NM, Serpa SS, Nunes JLS. Análise comparativa da pesca de curral na Ilha de São Luís, estado do Maranhão, Brasil. Arq Ciências do Mar. 2009;42(1):1–7.

[26] Almeida LGD. Caracterização das áreas de pesca artesanal de lagosta na Praia da Redonda, Icapuí-Ce. [dissertation]. Fortaleza: Federal University of Ceará; 2010.

[27] IBAMA. Instrução Normativa n 138, de 6 de dezembro de 2006. [Internet]. 2006. Available http://www.ibama.gov.br/documentos-recursos-pesqueiros/instrucao-normativa. [Accessed: 2016-01-12].

[28] IBAMA. Instrução Normativa n 206, de 14 de novembro de 2006. [Internet]. 2006. Available http://www.ibama.gov.br/documentos-recursos-pesqueiros/instrucao-normativa. [Accessed: 2016-01-12].

[29] Brasil. Lei n 9.605, de 12 de fevereiro de 1998. [Internet]. 1998. Available from: http://www.planalto.gov.br/ccivil_03/LEIS/L9605.htm. [Accessed: 2016-01-18].

[30] Silva MR. Povos de Terra e Água: a comunidade pesqueira de Canto do Mangue, Canguaretama (RN)- Brasil. [dissertation]. Piracicaba: University of São Paulo; 2004.

[31] Rangely J, Fabré NN, Tiburtino C, Batista V. Estratégias de pesca artesanal no litoral marinho alagoano (Brasil). Bol do Inst Pesca. 2010;36(4):263–75.

[32] Souza CD, Batista VS, Fabré NN. Caracterização da pesca no extremo sul da Área de Proteção Ambiental Costa dos Corais, Alagoas, Brasil. Bol do Inst Pesca. 2012;38(2):155–69.

[33] Burda CL, Schiavetti A. Análise ecológica da pesca artesanal em quatro comunidades pesqueiras da Costa de Itacaré, Bahia, Brasil: Subsídios para a Gestão. Rev da Gestão Costeira Integr. 2008;8(2):149–68. DOI: 10.5894/rgci136.

[34] Diegues ACS. Ecologia humana e planejamento em áreas costeiras. 2nd ed. São Paulo: NUPAUB-USP; 2001. 191 p.

[35] Castro J, editor. Homens e caranguejos. 2nd ed. Rio de Janeiro: Civilização Brasileira; 2001. 203 p.

[36] Brasil. Decreto n 6040, de 7 de fevereiro de 2007. [Internet]. 2007. Available from: https://www.planalto.gov.br/ccivil_03/_ato2007-2010/2007/decreto/d6040.htm. [Accessed: 2016-01-18].

[37] Diegues ACS. Povos e mares: leituras em sócio-antropologia marítima. São Paulo: NUPAUB-USP; 1995. 260 p.

[38] Mourão JS, Nordi N. Etnoictiologia de pescadores artesanais do estuário do Rio Mamanguape, Paraíba, Brasil. Bol do Inst Pesca. 2003;29(1):9–17.

[39] Condini MV, Garcia AM, Vieira JP. Descrição da pesca e perfil sócio-econômico do pescador da garoupa-verdadeira Epinephelus marginatus (Lowe) (Serranidae: Epinephelinae) no Molhe Oeste da Barra de Rio Grande, Rio Grande do Sul, Brasil. Panam J Aquat Sci. 2007;2(3):279–87.

[40] Fagundes L, Tomás ARG, Casarini LM, Bueno EF, Lopes GM, Machado DAL, Rosa RA, Braga ACA, Camargo FBF, Oberg IMF, Pellegrini SOP. Pesca de arrasto na Ilha de São Vicente, São Paulo, Brasil. Sér. Relat. Téc. 2007;29:1–43.

[41] Fuzetti L. A pesca na Ilha do Mel (Paraná-Brasil): Pescadores, atividades e recursos pesqueiros. [dissertation]. Curitiba: Federal University of Paraná; 2007.

[42] Souza KM, Arfelli CA, Lopes RG. Perfil socioeconômico dos pescadores de camarão-sete-barbas (*Xiphopenaeus kroyeri*) da praia do Perequê, Guarujá-SP. Bol do Inst Pesca. 2009;35(4):647–55.

[43] Mariz D, Souza ACFF, Teixeira SF, Campos SS, Lucena RFP, Alves RRN. Effects of urban development on socioeconomic aspects of a tropical artisanal fishing community. Indian J Tradit Knowl. 2014;13(4):637–46.

[44] Clauzet M, Ramires M, Begossi A. Etnoictiologia dos pescadores artesanais da praia de Guaibim, Valença (BA), Brasil. Neotrop Biol Conserv. 2007;2(3):136–54.

[45] Lira L, Mesquita B, Souza MMC, Leite CA, Leite APA, Farias AM, Galvão C. Diagnóstico socioeconômico da pesca artesanal do litoral de Pernambuco. Recife: Instituto Oceanário de Pernambuco e Departamento de Pesca e Aquicultura da UFRPE; 2010. 116 p.

[46] Vasconcellos M, Diegues AC, Kalikoski DC. Coastal fisheries of Brazil. In: Salas S, Chuenpagdee R, Charles A, Seijo JC, editors. Coastal fisheries of Latin America and the Caribbean. Rome: Food and Agriculture Organization of the United Nations; 2011. p. 73–116.

[47] IBGE. Mulheres mais escolarizadas são mães mais tarde e têm menos filhos. [Internet]. 2011. Available from:http://www.ibge.gov.br/home/presidencia/noticias/noticia_visualiza.php?id_noticia=1717&id_pagina=1. [Accessed: 2011-04-12].

[48] Alves RRN, Nishida AK. Aspectos socioeconômicos e percepção ambiental dos catadores de caranguejo-uçá *Ucides cordatus cordatus* (L. 1763) (Decapoda, Brachyura) do estuário do Rio Mamanguape, nordeste do Brasil. Interciencia. 2003;28(1):36–43.

[49] Nishida AK, Nordi N, Alves RRN, Aspectos socioeconômicos dos catadores de moluscos do litoral paraibano, Nordeste do Brasil. Rev Biol Ciênc Terra. 2008;8(1):207–15.

[50] IBGE. Pesquisa nacional de saneamento básico. [Internet]. 2002. Available from: http://www.ibge.gov.br/home/estatistica/populacao/condicaodevida/pnsb/pnsb.pdf. [Accessed: 2016-01-12].

[51] Diegues AC, org. Os saberes tradicionais e a biodiversidade no Brasil. São Paulo: MMA/COBIO/NUPAUB/USP; 2000. 211 p.

[52] Mariz D, Souza ACFF, Teixeira SF, Campos SS, Lucena RFP, et al. "Todo peixe no mar come e é comido": o discurso do sujeito coletivo sobre o uso de iscas pelos pescadores artesanais marinhos de Recife (Pernambuco, Brasil). Gaia Sci. (Special Edition). 2014;51–61.

[53] Carneiro M, Farrapeira C, Silva K. O manguezal na visão etnoecológica dos pescadores artesanais do Canal de Santa Cruz, Itapissuma, Pernambuco, Brasil. Biotemas. 2008;21(4):147–55. DOI: http://dx.doi.org/10.5007/2175-7925.2008v21n4p147.

[54] Amorim AN. Etnobiologia da comunidade de pescadores artesanais urbanos do bairro Poti Velho, Teresina/PI, Brasil [dissertation]. Salvador: Federal University of Bahia; 2010.

[55] El-Deir SG. Gestão Ambiental; I - Percepção ambiental e caracterização sócio-econômica e cultural da comunidade de Vila Velha, Itamaracá - PE (Brasil). Trab Ocean da Univ Fed Pernambuco. 1999;27(1):175–85.

[56] Pedrosa RA. Pesca, perfil socioeconômico e percepção ecológica dos pescadores artesanais de Porto de Galinhas (PE). [dissertation]. Recife: Federal University of Pernambuco; 2007.

[57] Araújo IX, Sassi R, Lima ERV. Pescadores artesanais e pressão imobiliária urbana: qual o destino dessas comunidades tradicionais? Rev Gestão Costeira Integr. 2014;14(3):429–46. DOI: 10.5894/rgci482.

[58] Evangelista-Barreto NS, Cleusa A, Daltro S, Paim I, Bernardes FDS. Indicadores socioeconômicos e percepção ambiental de pescadores em São Francisco do Conde, Bahia. Brazil. Bol Inst Pesca. 2014;40(3):459–70.

[59] Santos EC, Sampaio CLS. A pesca artesanal na comunidade de Fernão Velho, Maceió (Alagoas, Brasil): de tradicional a marginal. Rev Gestão Costeira Integr. 2013;13(4):513–24. DOI:10.5894/rgci428.

[60] Little PE. Territórios Sociais e Povos Tradicionais no Brasil: por uma antropologia da territorialidade. [Internet]. 2002. Available from: http://www.direito.mppr.mp.br/arquivos/File/PaulLittle.pdf. [Accessed: 2016-01-19].

[61] Sack RD. Human territoriality: Its theory and history. Cambridge: Cambridge University Press; 1986. 19 p.

[62] Cunha LHO. Saberes patrimoniais pesqueiros. Desenvolv e Meio Ambient. 2003;(7):69–76. DOI: http://dx.doi.org/10.5380/dma.v7i0.3044.

[63] Vasconcellos M, Diegues AC, Sales RR. Alguns aspectos relevantes relacionados à pesca artesanal costeira nacional. [Internet]. 2007. Available from: http://www. usp.br/nupaub/SEAPRelatorio.pdf. [Accessed: 2016-01-18].

[64] Marcelino RL, Sassi R, Cordeiro TA, Costa CF. Uma abordagem sócio-econômica e sócio-ambiental dos pescadores artesanais e outros usuários ribeirinhos do estuário do Rio Paraíba do Norte, Estado da Paraíba, Brasil. Trop Oceanogr. 2005;33(2):183–197.

[65] Rebouças GN, Filardi ACL, Vieira PF. Gestão integrada e participativa da pesca artesanal: potencialidades e obstáculos no litoral do Estado de Santa Catarina. Ambient Soc. 2006;9(2):83–104. DOI: http://dx.doi.org/10.1590/S1414-753X2006000200005.

[66] Souza PL, Coriolano LN, Ferreira MC, Alveirinho DJ. O Nordeste brasileiro e a Gestão Costeira. Rev Gestão Costeira Integr. 2008; 8(2):5–10. DOI: 10.5894/rgci58.

[67] Lacerda N. "Mundos" distintos: conflitos pela apropriação do litoral nordestino do brasil. Rev Bras Estud urbanos e Reg. 2010;12(2):39–52.

[68] Rabelo E. Brasília Teimosa. Edição Extra. Recife; 1968. ano1, n.2, p.10.

[69] Miranda L, Moraes D. Plano de regularização das Zonas Especiais de Interesse Social (PREZEIS) do Recife: democratização da gestão e planejamento participativo. In: Cardoso AL, coord. Habitação Social nas Metrópoles Brasileiras: uma avaliação das políticas habitacionais em Belém, Belo Horizonte, Porto Alegre, Recife, Rio de Janeiro e São Paulo no final do século XX. Porto Alegre: IPPUR; 2007. Coleção Habitare. p. 414–36.

[70] Carvalho-Souza GF, Tinôco MS. Avaliação do lixo marinho em costões rochosos na Baía de Todos os Santos, Bahia, Brasil. Rev Gestão Costeira Integr. 2011;11(1):135–43. DOI: 10.5894/rgci231.

[71] Pernambuco. RIMA - Relatório de impacto ambiental: Recuperação da Orla Marítima – Municípios de Jaboatão dos Guararapes, Recife, Olinda e Paulista (Pernambuco). [Internet]. 2012. Available from: http://www.cprh.pe.gov.br/downloads/rima_recuperacao_orla_maritima.pdf. [Accessed: 2016-01-19].

[72] Callou ABF. Povos do mar: herança sociocultural e perspectivas no Brasil. Cienc Cult. 2010;62(3):45–8.

[73] Loureiro VR. Os parceiros do mar: natureza e conflito social na pesca da Amazônia. Belém: CNPq e MPEG; 1985. 227 p.

[74] Mendonça SAT, Valencio NFLS. O papel da modernidade no rompimento da tradição: as políticas da SEAP como dissolução do modo de vida da pesca artesanal. Bol do Inst Pesca. 2008;34(1):107–16.

[75] Cardoso ES. Mundo do trabalho e pesca: apontamentos para a investigação. Rev Pegada. 2009;10(2):1–14.

[76] Ministério da Pesca e Aquicultura (MPA). Histórico [Internet]. 2015. Available from: http://www.mpa.gov.br/infraestrutura-e-fomento/116-ministerio/1406-historico [Accessed: 2015-11-12].

[77] Brasil. Lei n° 9.445, de 14 de março de 1997. [Internet]. 1997. Available from: http://www.planalto.gov.br/Ccivil_03/LEIS/L9445.htm. [Accessed: 2016-01-12].

[78] Brasil. Lei n° 10.683, de 28 de maio de 2003. [Internet]. 2003. Available from: http://www.planalto.gov.br/ccivil_03/leis/2003/L10.683compilado.htm. [Accessed: 2016-01-12].

[79] Brasil. Instrução Normativa n° 6, de 19 de maio de 2011. [Internet]. 2011. Available from: http://sinpesq.mpa.gov.br/rgp_cms/images/publico/CGRA/instrucao_normativa_06_de_19mai11-registro_e_licenca_aquicultor.doc. [Accessed: 2016-01-12].

[80] Brasil. Lei n° 10.779, de 25 de novembro de 2003. [Internet]. 2003. Available from: http://www.planalto.gov.br/ccivil_03/leis/2003/L10.779.htm. [Accessed: 2016-01-12].

[81] Brasil. Lei n° 11.326, de 24 de julho de 2006. [Internet]. 2006. Available from: http://www.planalto.gov.br/ccivil_03/_ato2004-2006/2006/lei/l11326.htm. [Accessed: 2016-01-12].

[82] Brasil. Lei n° 11.699, de 13 de junho de 2009. [Internet]. 2009. Available from: http://www.planalto.gov.br/ccivil_03/_ato2007-2010/2008/lei/l11699.htm. [Accessed: 2016-01-12].

[83] Haimovici M, Andriguetto Filho JM, Sunye PS, Martins AG. Padrões das dinâmicas de transformação em pescarias marinhas e estuarinas do Brasil (1960–2010). In: Haimovici M, Andriguetto Filho JM, Sunye PS, editors. A pesca marinha e estuarina no Brasil: estudos de caso multidisciplinares. Rio Grande: Editora da FURG; 2014. p. 181–91.

[84] Brasil. Medida Provisória n° 696, de 2 de outubro de 2015. [Internet]. 2015. Available from: http://www.planalto.gov.br/ccivil_03/_Ato2015-2018/2015/Mpv/mpv696.htm. [Accessed: 2016-01-12].

[85] FAO (Food and Agriculture Organization of the United Nations). Voluntary guidelines for securing sustainable small-scale fisheries in the context of food security and poverty eradication. Rome: FAO; 2015. 18 p.

[86] Pauly D, Christensen V, Guénette S, Pitcher TJ, Sumaila UR, Walters CJ, et al. Towards sustainability in world fisheries. Nature. 2002;418(8):689–95. DOI: 10.1038/nature01017.

[87] Schiermeier Q. Fisheries science: How many more fish in the sea? Nature. 2002;419:662–5. DOI: 10.1038/419662a.

[88] Myers RA, Worm B. Rapid worldwide depletion of predatory fish communities. Lett Nature. 2003;423: 280–3. DOI:10.1038/nature01610.

[89] Anticamara JA, Watson R, Gelchu A, Pauly D. Global fishing effort (1950–2010): trends, gaps, and implications. Fish Res. 2011;107(1–3): 131–6. DOI: 10.1016/j.fishres.2010.10.016

[90] Stewart KR, Lewison RL, Dunn DC, Bjorkland RH, Kelez S, Halpin PN, et al. Characterizing fishing effort and spatial extent of coastal fisheries. PLoS One. 2010;5(12):1–8. DOI: 10.1371/journal.pone.0014451

[91] Pauly D, Zeller D. Catch reconstructions reveal that global marine fisheries catches are higher than reported and declining. Nat Commun. 2016;1–9. DOI: 10.1038/ncomms10244

[92] FAO (Food and Agriculture Organization of the United Nations). Code of Conduct for Responsible Fisheries. Rome: FAO; 1995. 41 p.

[93] Garcia SMM, Zerbi A, Aliaume C, Do Chi T, Lasserre G. The ecosystem approach to fisheries. Issues, terminology, principles, institutional foundations, implementation and outlook. FAO Fisheries Technical Paper (no. 443). Rome: FAO; 2003. 71 p.

[94] Pitcher TJ. Rapfish, a rapid appraisal technique for fisheries, and its application to the code of conduct for responsible fisheries. FAO Fisheries Cicular (no. 947). Rome: FAO; 1999. 47 p.

[95] Pitcher TJ, Preikshot D. RAPFISH: A rapid appraisal technique to evaluate the sustainability status of fisheries. Fish Res. 2001;49:255–70. DOI: 10.1016/S0165-7836(00)00205-8.

[96] Haimovici M., editor. Sistemas pesqueiros marinhos e estuarinos do Brasil: caracterização e análise da sustentabilidade. Rio Grande: Editora da FURG; 2011. 104 p.

[97] FAO (Food and Agriculture Organization of the United Nations). Guidelines for the ecolabelling of fish and fishery products from marine capture fisheries - Revision 1. Rome: FAO; 2009. 90 p.

[98] ISEAL (International Social and Environmental Accreditation and Labelling Alliance). Our Codes of Good Practice [Internet]. Available from: http://www.isealalliance.org/our-work/defining-credibility/codes-of-good-practice [Accessed: 2016-01-23].

[99] MSC (Marine Stewardship Council). How we meet best practice [Internet]. Available from: https://www.msc.org/about-us/credibility/how-we-meet-best-practice?set_language=en [Accessed: 2016-01-23].

[100] Cummins A. The Marine Stewardship Council: a multi-stakeholder approach to sustainable fishing. Corp Soc Responsib Environ Manag. 2004;11(2):85–94. DOI: 10.1002/csr.56.

[101] Ponte S. The Marine Stewardship Council (MSC) and the making of a market for 'sustainable fish'. J Agrar Chang. 2012;12: 300–15. DOI: 10.1111/j.1471-0366.2011.00345.x

[102] Berkes F. Social aspects of fishery management. In: Cochrane, K. L., and Garcia, S. M., editors. A Fishery Manager's Guidebook. Singapore: The Food and Agriculture Organization of the United Nations and Wiley-Blackwell; 2009. p. 52–74.

[103] Jackson JB., Kirby MX, Berger WH, Bjorndal KA, Botsford LW, Bourque BJ, et al. Historical overfishing and the recent collapse of coastal ecosystems. Science. 2001;293(5530): 629–637. DOI: 10.1126/science.1059199.

[104] Gelpke N, Behnam A, Visbeck M, editors. World Ocean Review 2013: Living with the oceans. 2: The future of fish – the fisheries of the future. Hamburg: Maribus gGmbH; 2013. 143 p.

[105] Fernandez, JIV. A política nacional de desenvolvimento sustentável da pesca e da aquicultura e seus impactos sobre a pesca artesanal no Estado de Pernambuco [dissertation]. Recife: Universidade Federal de Pernambuco; 2015.

[106] Pedrosa BMJ, Lira L, Santiago ALM. Pescadores urbanos da zona costeira do estado de Pernambuco, Brasil. Bol do Inst Pesca. 2013;39(2):93–106.

[107] Sanders JS, Greboval D, Hjort A. Marine protected areas: country case studies on policy, governance and institutional issues. Rome: FAO; 2011. 118 p.

[108] Cordell J. Dynamics and challenges of MPA development and coastal protection. In: Taking marine management to scale: connecting societies, coastal landscapes and the sea. Washington: World Bank, 2006.

[109] Diegues AC. Marine Protected Areas and Artisanal Fisheries in Brazil. Chennai: International Collective in Support of Fishworkers (ICSF); 2008. 54 p.

9

Sustainable Urbanization in the China-Indochinese Peninsula Economic Corridor

Dong Jiang, Jingying Fu and Gang Lin

Abstract

Countries in the China-Indochinese Peninsula are home to rich human and natural resource endowments and have the potential to be one of the world's fastest growing areas. Sustainable urbanization in the China-Indochinese Peninsula Economic Corridor is important for the regional economic development and prosperity. Taking the advantages of the remote sensing and Geographic Information System (GIS) technologies, this chapter is first presents a general overview of urbanization procession in this region and monitors the spatiotemporal dynamics of the urban environment; the second objective is to present the multiple driving force factor analysis for urban development in countries of the China-Indochinese Peninsula Economic Corridor using statistical models. The results indicated that the China-Indochinese Peninsula Economic Corridor has experienced a rapid urbanization process during the past 15 years both in terms of urban areas and urban population (UP). In addition to socioeconomic factors, there is also a noticeable correlation between foreign direct investment (FDI) and international trade and urban development in the China-Indochinese Peninsula Economic Corridor. Active participation in international trade and attracting foreign investment are helpful for the regional urbanization. As a neighboring country, China's economic and trade activity also has a significant impact on the urbanization in countries of the China-Indochinese Peninsula Economic Corridor. Furthermore, as the launch of the Silk Road Economic Belt and the 21st Century Maritime Silk Road and the Asian Infrastructure Investment Bank (AIIB), the China-Indochinese Peninsula Economic Corridor will witness a more rapid urbanization progress in the next decade. This study has its characteristics in focusing on the region of the Indochinese Peninsula in which the most rapid urbanization is occurring, presenting the state-of-the-art techniques for monitoring urban expansion and probing into the driving factors of the urban expansion in the China-Indochinese Peninsula Economic Corridor by multiple principles and multiple-level data. It is expected to benefit policymakers in urban development and also provide a basis for further studies of sustainable urbanization in the China-Indochinese Peninsula Economic Corridor.

Keywords: sustainable urbanization, China-Indochinese Peninsula Economic Corridor, remote sensing, GIS, driving force analysis

1. Introduction

Urbanization is one of the most powerful and visible anthropogenic forces on Earth. Although urban areas only occupy a relatively small part of the Earth's land area, they represent 54% of the global population (and even more in the following decades) [1]. With rapid economic globalization, urbanization is now having a huge impact on the political, socioeconomic, and environmental landscape of countries across the world. In recent years, taking advantage of remote sensing, many studies have been performed by scholars from universities, academic institutions, and international organizations on different subjects related to urbanization. Funded by the National Aeronautics and Space Administration (NASA), the 100 Cities Project was implemented in 2010 by Arizona State University (ASU) to supply remote sensing images of 100 international cities as a tool for creating urban models and formulating an effective policy for policymakers and researchers from around the world. The data set generated by this project could be used to create sustainable urban planning practices in various climatic, ecological, and social regions [2]. In 2012, using the urban and rural information derived from satellite data and other sources, NASA's Socioeconomic Data and Applications Center (SEDAC) launched the Global Rural-Urban Mapping Project to respond to the challenges of sustainable development and environmental management presented by world urbanization. That project presented a series of spatial distribution data of human populations to study urban ecology and address critical environmental and societal issues in urban areas [3]. Recently, the World Bank (WB), in collaboration with the University of Wisconsin and the WorldPop project, has developed a map of built-up areas, urban expansion, and urban population (UP) changes across the East Asian region (stretching from Mongolia to the Pacific Islands) for the years 2000 and 2010. These data sets include data on all 869 urban areas in the region with populations of more than 100,000 and serve as a valuable reference for urban geography studies on changing patterns of urbanization [4]. Meanwhile, many scholars also devote urban development studies using remote sensing technology [5, 6], and studies on sustainable urbanization are ongoing. To leave extreme poverty behind and prosper, East-Southeast Asia is currently experiencing rapid urbanization, and cities play a transformative role in this economic growth. Sustainable urban development in the cities of East-Southeast Asia draws an increasing amount of global attention to the region's stability and development.

Figure 1. Location of the Indochinese Peninsula.

The Indochinese Peninsula is located between China and the South Asian Subcontinent. It is bordered by the Bay of Bengal, the Andaman Sea and Malacca in the west and the South China Sea in the east (**Figure 1**). In the Indochinese Peninsula, a peninsula in southeastern Asia that contains Myanmar, Cambodia, Laos, Thailand, and Vietnam, the rapid urbanization of recent years exerted strong influences on regional development. In 1992, with assistance from the Asian Development Bank (ADB), those five countries and China established a program of subregional economic cooperation in the Greater Mekong Subregion (GMS) that aimed to enhance economic relations among them. The GMS program helps the five countries implement many high-priority subregional projects in transportation, energy, telecommunications, the environment, human resource development, tourism, trade, private sector investment, and agriculture, all of which were strong drivers of the economy and urbanization process in the Indochinese Peninsula. **Table 1** shows the number of cites in the Indochinese Peninsula in 2015 classified by country-size class; that information was obtained from the World Urbanization Prospects reported by the United Nations (UN) [1]. In the five countries of the Indochinese Peninsula, there are 26 cities with populations of more than 300,000 and 16 cities with populations of more than 500,000. From the table, we conclude that Thailand and Vietnam contain many more big urban agglomerations than the other countries, thus showing their high level of urbanization. Second, despite the presence of some metropolitan cities, urbanization in countries of the Indochinese Peninsula is broad-based. Only 1 city with a population of more than 300,000 can be found in Cambodia and Laos; Thailand and Vietnam have only 1 supercity (5 million or more inhabitants) each. Increasingly, after a process of transition and transformation, modernization and industrialization are emerging in Myanmar, Cambodia, Laos, Thailand, and Vietnam. These countries are gradually shifting from traditional farming to more diversified economies and to more open market-based systems [7]. Parallel with this development are the growing economy links between the five countries and their neighbors, notably in terms of cross-border trade, investment, and labor mobility [8]. Moreover, natural resources, particularly hydropower, are beginning to be developed and used in the region [9]. The Silk Road Economic Belt and the 21st Century Maritime Silk Road [One Belt and One Road (B&R)], which was proposed by China in 2014, comprise a development strategy and framework that aims to enhance the economic relationship among countries in Asia and Europe [10]. The China-Indochinese Peninsula Economic Corridor, as an important international gateway of B&R, is supposed to develop a regional economic entity with common development that uses the railways and roads as a medium. The rich human and natural resource endowments of the Indochinese Peninsula region have made it a new frontier of Asian economic growth. The Indochinese Peninsula has tremendous potential to promote both regional economic growth and urban development. Thus, following the 2014 launch of B&R, the China-Indochinese Peninsula Economic Corridor will witness a more rapid urbanization progress in the next decade.

Size class	Cambodia	Laos	Myanmar	Thailand	Vietnam
5 million or more	–	–	–	1	1
1–5 million	1	–	3	1	3
500,000–1 million	–	1	1	2	2
300,000–500,000	–	–	2	5	3

Sources: World Urbanization Prospects: The 2014 Revision, UN.

Table 1. Number of agglomerations classified by country-size class in the Indochinese Peninsula, 2015.

This chapter gives a general overview of the urbanization procession in countries in the Indochinese Peninsula region and presents the state-of-the-art techniques for monitoring the spatiotemporal dynamics of the urban environment. An analysis of the forces driving urban expansion was also performed based on an integrated analysis of both natural and social economic factors.

2. Methodology

2.1. Monitoring urban expansion in the Indochinese Peninsula

Urban sprawl monitoring constitutes basic information for urban studies, and accurate information about the extent of urban growth is of great interest to researchers investigating urbanization progress. Long ago, conventional surveying and mapping techniques were primarily used to estimate urban sprawl. These methods were usually expensive and time-consuming, and some key information was unavailable for most cities, especially in developing countries [11]. Fortunately, because remote sensing enjoys the advantages of being both cost-effective and technologically sound, it is increasingly used in the analysis of spatiography and urban geography. Since the 1970s, a variety of studies have been conducted using remote sensing and Geographic Information System (GIS) technologies to examine land-use change, to analyze large landscapes, to analyze farmland change and classification, and to analyze urban space structure and fractal shapes [12–15]. In recent years, with rapid economic development and population increases, rapid urbanization occurs and the city quickly expands; the strained relationship between the population and urban land-use is attracting increasingly broad scholarly attention. Thus, extensive research studies have been performed to monitor urban sprawl using remotely sensed images through either an image-to-image comparison or a postclassification comparison [16–19]. In the countries of the Indochinese Peninsula, there has also been a great deal of research effort devoted to land-use changes using remote sensing and GIS technologies. Kong et al. [20] investigated forestland changes attributed to urbanization and agricultural land expansion in Naypyidaw (Myanmar's capital) using Landsat images. Similarly, based on an analysis of the pattern of urban growth from 1993 to 2011 in Siem Reap, Cambodia, Ourng and Rodrigues [21] reported that urban growth always came accompanied primary roads and the river. Using remote sensing images, Okamoto et al. [22] studied the urbanization of Vientiane (capital of Laos). Kimijiama and Nagai [23] also presented the relationship between urbanization and socioeconomic activities in Savannaket, Laos. Nevertheless, because of the limited availability of regional spatial data, those previous studies rarely focused on urban sprawl at the national level; moreover, there is no systematic study available on the multiple temporal phases and long time series of urban sprawl monitoring by multiple-sourced remote sensing data in the countries of the Indochinese Peninsula. Remote sensing and GIS technologies have broad application space in the estimation of urban sprawl in this region.

2.1.1. Methods for extracting urban land with remote sensing

Remote sensing image classification is a primary method of extracting land surface information at large scale. Different ground objects have different spectral characteristics, which are recorded in satellite remote sensing images, and pixels with similar spectral characters would be considered as one landscape class.

The process of landscape classification is complex. Its accuracy is always influenced by many factors such as the sources of the data, the image quality in remotely sensed images, and the method selected for the classification [24]. The classification method is important for landscape classification. Typically, the classification can be divided into two types: (1) pixels compared one by one, which involves monitoring the changes to each pixel by comparing the various temporal phases, and (2) first classifying and then comparing, which involves classifying the remote sensing images in different temporal phases separately and then comparing the classification results to monitor the land-use change [25, 26]. The various methods all have their own advantages and disadvantages; therefore, there may be no single method that is suitable for all situations. For example, although the previous method is simple and easy to achieve, it only monitors the changes of the pixel rather than obtaining the changes of the objects; the second method is limited by classification accuracy in different temporal phases and accumulates calculated errors, thus making its accuracy dependent on the accuracy of the previous classification result. In reality, we should select a landscape classification method based on our needs. The common methods for extracting urban information are as follows:

2.1.1.1. Supervised classification

Supervised classification is also known as the training classification method. In this method, we select training samples of different land-use types in the image and analyze the sample information for each training area by the computer. Next, each pixel can be classified into the similar sample area based on the comparative result of the pixel and the training samples [27–29]. The common methods for supervised classification are the single linkage method [30], the Fisher discriminant method [31, 32], the Bayes linear discriminant analysis [33], and the maximum likelihood method [34]. Supervised classification has the advantage of selectively determining the quantitative and categorical classification based on the study object and area while eliminating needless classifications. Additionally, supervised classification can control the selection of the training samples. Supervised classification is limited, however, by the subjective factors of humans to select the training samples and determine the classification system [35].

2.1.1.2. Unsupervised classification

Unsupervised classification primarily relies on the structural features of the image data and natural points for the object classification without the known training data and the number of the classification. Based on similar levels of the luminance value of the samples in the multi-dimensional spectrum space, the computer can automatically analyze the classifying parameters and then classify the pixels accordingly [36–38].

Unsupervised classification depends on similar levels of the pixels' luminance value for object classification instead of relying on prior knowledge. The dependency on spectra quality replicates the biggest flaw of unsupervised classification; because of differences in location, shape, and character, the same ground objects may have different manifestations in the spectra image, inducing errors in classification. Compared with the method of supervised classification, the unsupervised classification is neither fast nor precise but does have the advantage of being highly objective [39]. According to the study by Xue and Ni [40], although unsupervised classification is unsuitable for the extraction of residential areas compared to the supervised classification, the selection of the training areas has a greater impact on classification accuracy determined by the supervised classification method. Moreover, Hu et al. [41] also extracted urban land-use information using the two methods.

2.1.1.3. Visual interpretation

Visual interpretation relates to extracting information about specified ground objects from the remote sensing image using either direct observations or assisted instruments [42]. Visual interpretation, which enjoys the advantages of easy operation and less equipment, is popular with geographers. The interpretation marks of visual interpretation can be categorized as direct and indirect. The ground object directly interpreted by the characteristic of the image is defined as the interpretation mark. The basic elements of the direct interpretation mark are the tone and the diagram: the tone reflects the image's physical property, whereas the diagram reflects its geometric properties. The tone is the analog recording of the grayscale, which shows the color code and chroma in the color image. In visual interpretation, although the tone of the recognized ground object is a quick mark, it is an uncertainty criterion because it suffers from various influence factors. Thus, the tone could merely be a relative reference for interpretation, and we cannot identify the ground object by relying exclusively on it [43]. The diagram can reflect the shape, size, location, and plane relation among the ground objects. In general, geomorphologic shape, vegetation distribution, water bodies, and bare land are all interpreted by the tone and diagram. Moreover, clouds, snow, urban land, open-pit mines, and airports can be identified by the image [44]. Based on the studies of the phenomenon close to the interpreted objects, indirect interpretation is defined as inferring and distinguishing among the ground objects. Location, relative positions, and other things close to the interpreted objects can all be regarded as indirect marks. Location is both the reflection of the environment of the objects in the image and the relationship between the objects and the environment. Relative positions relate to the plane layout of the dependence relationship among the landscape elements and the objects in the image [43]. The common methods of visual interpretation include the direct identification method, the comparison method, and logical reasoning. The direct identification method can quickly interpret ground objects by their marks, whereas logical reasoning needs to identify objects' existence and properties by the appearance of the internal relations of the objects or natural phenomena. The comparison method first compares the ground objects and natural phenomenon in the image with a known remote sensing image and then identifies the properties of the objects. In any event, it is very important to analyze the comprehensive feature of interpretation objects for each visual interpretation method. To improve the precision of the interpretation, direct and indirect interpretation marks should be used conjunctively, and the image should be taken as a contrastive analysis of various bands and temporal phases [45].

2.1.1.4. Automatic classification and change detection

On a regional scale, moderate spatial resolution remote sensing images such as land resource satellite data are usually used for the data used in landscape classification and change detection. To rapidly achieve an accurate classification rapidly and to minimize human intervention, Jiang et al. propose an efficient, automatic landscape classification approach taking prior accurate land-cover data as the background experience [46, 47]. By adopting the prior knowledge, this approach is distinguished from the previous semisupervised findings of landscape classifiers. This approach involves two steps. First, based on the historical image data, one detects changed landscape pixels from satellite images. Second, one classifies the changed pixels in the landscape based on pattern recognition and changed rules. This approach enjoys the advantages of multimethods in landscape classification, primarily described as follows: (1) the historical data for land-use cover is high precision and can be better matched with the remote sensing data, and (2) based on the ecology view, this approach assumes that the junction of different land types is the fragile area, which is the main changed areas and the inner area is the relatively stable region. Furthermore, the big plaque will be more stable than the small one. This approach can be applied not only to microsatellite data but also to landscape classification for other spectral remote sensing images.

2.1.1.5. Normalized Difference Built-up Index (NDBI) method

Based on the deep analysis of the Normalized Difference Vegetation Index (NDVI), the NDBI was first proposed by Yang [48] and later improved by Zha and Ni [39]. For the Landsat TM image, the gray value of the objects will show only small changes, except for the urban land in the bands of TM4 and TM5. Based on the spectral characteristics, NDBI achieves urban land extraction using the following formula:

$$NDBI = \frac{(TM5 - TM4)}{(TM5 + TM4)} \tag{1}$$

where TM4 is band 4 and TM5 is band 5 of the TM image. Obviously, the value of NDBI should be between −1 and 1; after the two binary transform, the interval value of −1 and 0 is assigned as 0 and the others are assigned as 255, then obtaining the binary image for urban land extracting.

Based on the National Oceanic and Atmospheric Administration images, NDVI, which is used for vegetation-information extraction on a regional scale, has tended to be mature. In recent years, this approach, which has been improved for urban land extracting, has been widely used in urban sprawl monitoring [41, 49].

2.1.1.6. Artificial neural network (ANN) classification

ANN is a complicated network system that is composed of abundant and simple processing elements; it contains engineering systems that simulate the operative mechanism and organizational structure of the human brain based on studies of human brain. ANN belongs to the nonparametric classifier. Since it was proposed in 1988, this approach has been attracting increasing attention to landscape classification and change detection. ANN is widely used in landscape classification such as land-use classification, ground object identification in different temporal phases, fuzzy classification, remote sensing image classification, and the extraction of the shape structure of the image. With the development of the theory of ANN and technological improvements, ANN has been an effective means for remote sensing classification. In recent years, numerous studies on ANN have been successfully performed for the geologic application. According to the characteristic of remote sensing, Dong et al. propose a landscape classification model based on the Hopfield ANN; this classifier proved to have better accuracy and higher efficiency than other methods [50]. Chen et al. have developed the Self Organizing Feature Map (SOFM) neural network, which is based on the weight of samples and data, to achieve the direct classification change detection for temporal and multispectral remote sensing data [51]. Common ANN models for landscape classification are the Multilayer Perceptron (MLP) classification model, the radial basis function (RBF) neural network classification model, the SOFM classification model, and the Adaptive Resonance Theory (ART) classification model. Aside from the above models, more other ANN models have been gradually applied to remote sensing classifications. In general, compared to other classification methods, ANN for land-use extraction has the following advantages: (1) it has the properties of self-learning, self-organizing, and self-adapting and can not only make maximum use of prior knowledge of the known samples type but also automatically extract the rules for multiclassification; (2) ANN need not make the assumptions of the probabilistic models; (3) with its capacity for fault tolerance, nonlinear decision boundaries can be developed in feature space in the ANN model; and (4) the ANN model has superior association power [52].

2.1.2. Satellite data for extracting urban land

2.1.2.1. Resource satellite data

Land resource satellites are primarily used for resource exploration and studying the natural ecoenvironment status of land surface; they are widely used for resource investigation, environmental monitoring, hazard monitoring, land-use planning, and regional development. The most common resource satellites are NASA's Landsat, France's SPOT, the China Brazil Earth Resource Satellite, India's Cartosat-1 (IRS-P5), and the Moderate Resolution Imaging Spectroradiometer (MODIS). The following is a detailed introduction to NASA's Landsat.

2.1.2.1.1. Landsat

Landsat-1, which the United States developed in 1972, is the first resource satellite in the world. It orbits at 704 km high and an angle of 98.2°; it circles the Earth in 16 days. Landsat-2 and Landsat-3 were launched in 1975 and 1978, respectively. The three satellites were the first-generation test satellites and carried the same sensors [i.e., the Return Beam Vidicon (RBV) and Multispectral Scanner System (MSS)]. Landsat-4 and Landsat-5 were launched in 1982 and 1984, respectively. They were first-generation practical satellites, carrying the MSS and Thematic Mapper (TM) sensors. Landsat-7, launched in 1999, was the third-generation satellite. The Enhanced TM Plus (ETM+) sensor, which had eight bands, was first carried in Landsat-7, replacing the TM sensor [53]. The new sensor can work for the spectral region of visible light, near infrared, shortwave infrared, and thermal infrared, and it contains the following improvements over the TM: (1) it introduced the panchromatic band with a resolution of 15 m, (2) the resolution of thermal infrared band was improved to 60 m, and (3) the solar calibrators reduced the satellite's radiation-calibration errors to less than 5%, which is five times that of a traditional satellite. Landsat-8, which carried the Operational Land Imager (OLI) sensor and the Thermal Infrared Sensor (TIRS), was launched in 2013 by NASA [54]. OLI includes all of the bands in ETM+, improved to exclude water absorption characteristics. Comparing the previous sensors, OLI excluded the water absorption characteristics in 0.825 μm in band 5; the range of panchromatic band 8 was narrow, which can help in differentiating the vegetation and nonvegetation. Furthermore, two added bands were included in Landsat-8, which were band 1 coastal (0.433–0.453 μm) for coastal zone monitoring and band 9 cirrus (1.360–1.390 μm) for clouds monitoring; the near- and short-infrared bands were closer to the MODIS bands. **Tables 2** and **3** show the Landsat launched by the United States and the bands included in ETM+ and OLI.

Satellite	**Landsat-1**	**Landsat-2**	**Landsat-3**	**Landsat-4**	**Landsat-5**	**Landsat-6**	**Landsat-7**	**Landsat-8**
Height (km)	920	920	920	705	705	–	705	705
Cycle (days)	18	18	18	16	16	16	16	16
Scan width (km)	185	185	185	185	185	185	185*170	170*180
Number of band	4	4	4	7	7	8	8	11
Sensor	MSS	MSS	MSS	MSS, TM	MSS, TM	ETM+	ETM+	OLI, TIRS

Table 2. Landsat launched by the United States for remote sensing monitoring.

Name	**OLI**		**Name**	**ETM+**	
	Band (μm)	**Spatial resolution (m)**		**Band (μm)**	**Spatial resolution (m)**
Band 1 Coastal	0.433–0.453	30			
Band 2 Blue	0.450–0.515	30	Band 1 Blue	0.450–0.515	30
Band 3 Green	0.525–0.600	30	Band 2 Green	0.525–0.605	30
Band 4 Red	0.630–0.680	30	Band 3 Red	0.630–0.690	30
Band 5 NIR	0.845–0.885	30	Band 4 NIR	0.775–0.900	30
Band 6 SWIR 1	1.560–1.660	30	Band 5 SWIR 1	1.550–1.750	30
Band 7 SWIR 2	2.100–2.300	30	Band 6 SWIR 2	2.090–2.350	30
Band 8 Pan	0.500–0.680	15	Band 7 Pan	0.520–0.900	15
Band 9 Cirrus	1.360–1.390	30			

Table 3. Bands included in ETM+ and OLI.

2.1.2.2. High-resolution satellite data

2.1.2.2.1. High spatial resolution remote sensing satellite

High spatial resolution remote sensing data have been a fundamental and strategic national resource, serving to provide accurate mapping, urban planning, land resource management, environmental monitoring, ground mapping, military mapping, and intelligence gathering. Because of its huge economic and military benefits, high spatial resolution remote sensing satellites are quickly developing all over the world. The IKONOS satellite, which was launched in 1999, marked the beginning of the commercial high-resolution satellite era; in 2001, the QuickBird satellite was developed by Digital Globe Company. **Table 4** shows the high spatial resolution remote sensing satellites that have a resolution of no more than 1 m [55].

Satellite	State	Launch time	Panchromatic resolution	Multispectrum resolution
OrbView 5	USA	2007	0.41	1.64
WorldView	USA	2006	0.5	2
QuickBird	USA	2001	0.6	2.5
EROS-B	Israel	2006	0.7	3.5
EROS-C	Israel	2008	0.7	2.5
Pleiades-1	France	2008	0.7	2.8
Pleiades-2	France	2009	0.7	2.8
IKONOS-2	USA	1999	1.0	4
OrbView 3	USA	2003	1.0	4
Kompsat	Korea	2004	1.0	4
Resurs DK	Russia	2005	1.0	3
IRS Cartosat 2	India	2006	1.0	–

Table 4. Bands included in ETM+ and OLI.

2.1.2.2.2. Hyperspectral remote sensing satellite

Hyperspectral resolution remote sensing is performed to obtain many narrow- and continuous-spectrum remote sensing images in the visible light, near infrared, intermediate infrared, and thermal infrared of the electromagnetic spectrum [56]. Hyperspectral resolution remote sensing technology contains the special properties of a fine structure of spectra and abundant date information; moreover, it has incomparable application advantages with respect to the identification, assortment, and information extraction of ground objects, which give it great potential for application to ecological environment monitoring [57].

On August 23, 1997, the first hyperspectral remote sensing satellite (LEWIS) was launched in the United States. After years of development, many hyperspectral remote sensing satellites have been developed and successfully operated. **Table 5** shows some of the hyperspectral resolution remote sensing satellites [57].

Satellites	Launch time	State	Band (nm)	Number of bands	Spatial resolution (m)
MODIS	1999	USA	400–1400	36	250, 500, 1000
ASTER	1999	USA–Japanese	520–860	3	15
			1600–2430	6	30
			8125–11,650	5	90
MightySat	2000	USA	500–1050	256 or 512	30
HYPERION	2000	USA	400–1100	60	30
			900–2500	160	
CHRIS	2001	European Space Agency (ESA)	400–1050	18–62	17
ARIES	2004	Australian	400–1100	60	30
			900–2500	160	
HSI	2008	Chian	450–950	115	100

Table 5. Common hyperspectral resolution remote sensing satellites.

2.1.3. Extracting urban land areas using remote sensing in the Indochinese Peninsula

2.1.3.1. Remote sensing data sources

In this chapter, for the study of urban expansion of the primary cities in the Indochinese Peninsula, Landsat TM/ETM+ and Landsat-8 images from 2000 to 2015 were primarily selected for urban area identification. Dynamic changes were analyzed using results from multiple years. In the process of interpreting the remote sensing data, Google Maps is an important reference for this region. **Table 6** shows the various types of high-quality satellite remote sensing data used in this study. To study urban expansion at the national level, spatial data on built-up areas in the East Asia region for the period from 2001 to 2010 were also used in this chapter.

2.1.3.2. Image processing of remote sensing

To study urban development in the Indochinese Peninsula, eight primary cities with populations of more than 500,000 were selected in this chapter: Naypyidaw and Yangon (Myanmar), Hanoi and Bien Hoa (Vietnam), and Bangkok and Chon Bury (Thailand) for the period from 2000 to 2015 and Vientiane (Laos) and Phnom Penh (Cambodia) for the period from 2000 to 2010 because the remote sensing data for Vientiane and Phnom Penh were deficient in 2015 (**Figure 1**). In this study, we extracted the urban land information from the remote sensing

Type of data	Years	Sources	Band
Landsat TM/ETM+	2000 and 2010	Global Land Cover Facility Earth Science Data Interface (URL: http://glcfapp.glcf.umd.edu:8080/esdi/index.jsp)	7, 4, 3
Landsat-8	2015	Download system for Landsat-8 (Chinese Academy of Sciences; URL: http://ids.ceode.ac.cn/query.html) China Centre For Resources Satellite Data and Application (URL: http://218.247.138.121/DSSPlatform/index.html)	7, 4, 3

Table 6. Types of high-quality satellite remote sensing data used in this study.

image by visual interpretation. This approach was used because of its advantages of simplicity and accuracy, although it is also time-consuming and costly. According to the shape and image features of the ground objects, most of the study area can be identified using this approach. For example, farmlands, water bodies, residential blocks, etc., can be easily recognized. Remote sensing TM7 is a medium-infrared waveband in which the rock shows a strong reflection, TM4 is a near-infrared waveband in which vegetation can be strongly reflected, and TM3 is the red waveband that shows the primary absorption of vegetation chlorophyll. Thus, we selected band combinations 7, 4, and 3, which can be used to identify the urban area with the characteristic of the built-up areas on less vegetation biota, whereas the suburban area shows abundant vegetation biota. **Figure 2** shows the technical route for the built-up area extraction, and **Figure 3a–d** shows the distribution of built-up areas in the representative cities of Yangon, Chon Bury, Bangkok, and Hanoi during various periods.

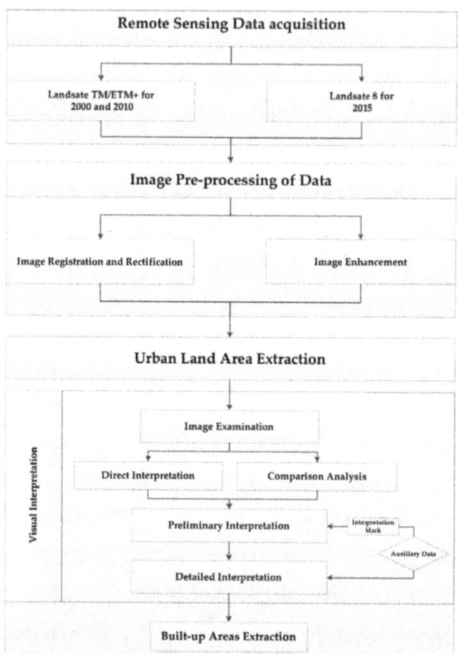

Figure 2. Map showing the technical route for the built-up area extraction.

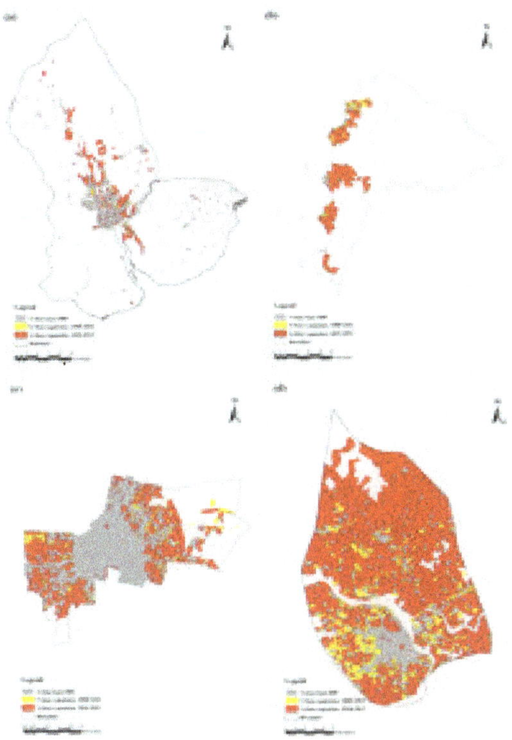

Figure 3. Maps showing the distribution of built-up areas in the cities of Yangon **(a)**, Chon Bury **(b)**, Hanoi **(c)**, and Bangkok **(d)** in the Indochinese Peninsula in different periods.

2.1.4. Urban expansion rate

In some of those previous studies [58, 59], the built-up area is considered an indicator for urban sprawl monitoring, and these areas always represent the status of a city's construction and development from the perspective of space in urban geography. Thus, in this study, the urban expansion rate was adopted to evaluate the spatial distribution and rate of urban sprawl in the Indochinese Peninsula for the period 2000 to 2015. The urban expansion rate that can be defined in Equation (2) shows changes in the quantity of the urban area per unit time and is a key parameter for evaluating spatial changes in urban sprawl [59, 60].

$$R_{UL} = \frac{UL_{n+i} - UL_i}{UL_i} \times \frac{1}{n} \times 100\% \quad (2)$$

where R_{UL} stands for the expansion rate of urban land; UL_{n+i} and UL_n stand for the built-up area in the target unit at times $n+i$ and i, respectively; and n is the interval of the calculation period (in years).

2.2. Methodology for driving force analysis for urban expansion

The dynamic changes of urban areas meet socioeconomic development, along with land use in the urban fringe area and the interior region, after continuous adjustment and configuration result in a transformation into urban land. With increased population, an increasing amount of the rural population is changed into an UP [61]. The dynamic changes of urban areas express urbanization in space and are an inevitable consequence of urbanization. Pattern-process-mechanism always guides the geographical study, and pattern is the distribution of the geographical objects and phenomena; process stands for the analysis of changes in the geographical objects and phenomena in time and space; and mechanism finds the reasons for these changes. Thus, driving force analysis for urban expansion can enable a better understanding of urban development and policy decisions [62]. This chapter presents a multiple-factor model (geographic position, regional economic development, population, infrastructure, and foreign economic and trade relations) to explore the driving forces of urban expansion in countries of the Indochinese Peninsula using multiple principles and multiple-level data.

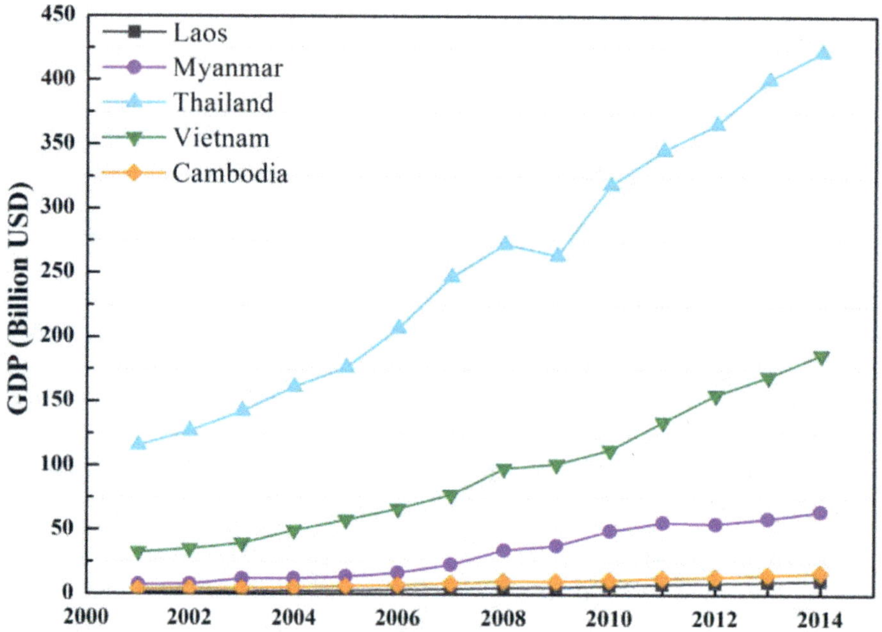

Figure 4. Map showing GDP in the countries of the Indochinese Peninsula.

2.2.1. Data sources for driving force analysis

2.2.1.1. Regional economic development

Urban development primarily rests on financial strength, and economic development accelerates city changes and urban expansion. To an extent, urban land-use can be viewed as an economic issue, which is also noted in prior studies [63, 64]. Thus, Gross Domestic Product (GDP) can be regarded both as an integrated index reflecting regional economic development and as a predictive factor for urban development. **Figure 4** shows GDP in the countries of the Indochinese Peninsula. These data were obtained from the GMS Statistics data set on the ADB Website, and the data set provided the latest state and trend of key GMS economic data according to the International Monetary Fund (IMF) World Economic Outlook [65, 66]. From

2000 to 2014, the economies of the countries of the Indochinese Peninsula have experienced rapid growth. Except for Thailand, the peninsula's economy has grown almost 7-fold over the last 15 years in Myanmar, Laos, Cambodia, and Vietnam.

2.2.1.2. Population data

To analyze the driving force of urban expansion, data on the UP (% of total) were used in this study. The number of people living in the urban land area is generally defined as UP, and the ratio of UP to total population relates to the percentage of the total population living in cities; UP (% of total) is usually regarded as an indicator of urbanization additional to built-up areas [67, 68]. The UP (% of total) data set for countries in the Indochinese Peninsula for the period from 2001 to 2014 used in this chapter were obtained from the WB's World Development Indicators (WDI), and these data show the numbers of urban residents per 100 total population [69]. The UN Department of Economic and Social Affairs, using the cohort component method, has developed population estimates for developing countries that lack census data. The Department calculated this data set and provided information that is convenient for population studies. These data are considered a valuable scientific reference for population studies, although there is some uncertainty caused by data limitations. **Figure 5** shows the UP (% of total) for countries in the Indochinese Peninsula for the period from 2001 to 2014. According to that figure, the UP proportion increased during the 14 years and the region appeared to experience rapid urbanization, especially in Thailand.

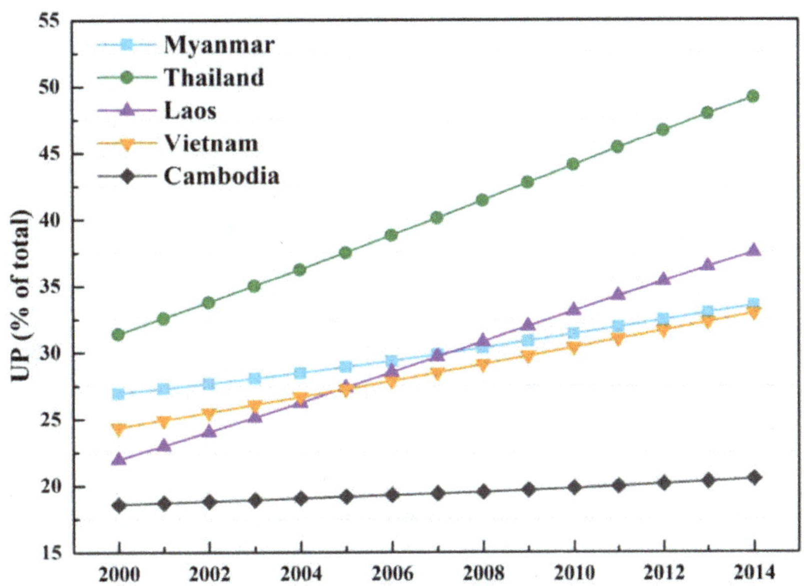

Figure 5. Map showing the UP (% of total) for countries in the Indochinese Peninsula for the period from 2001 to 2014.

To explore the relationship between urban expansion and UP, the UN's population statistics for urban agglomerations with 300,000 inhabitants or more in 2014 by country were used in this chapter to perform an urban expansion analysis of the eight cities in the Indochinese Peninsula [1]. Based on national statistics data (population censuses are the most commonly used sources), the UN developed the UP estimation to respond to the sustainable development challenges of urbanization. **Figure 6** shows the UP in Naypyidaw, Yangon, Hanoi, Bien Hoa, Bangkok, Chon Bury, Vientiane, and Phnom Penh in 2000, 2010, and 2015.

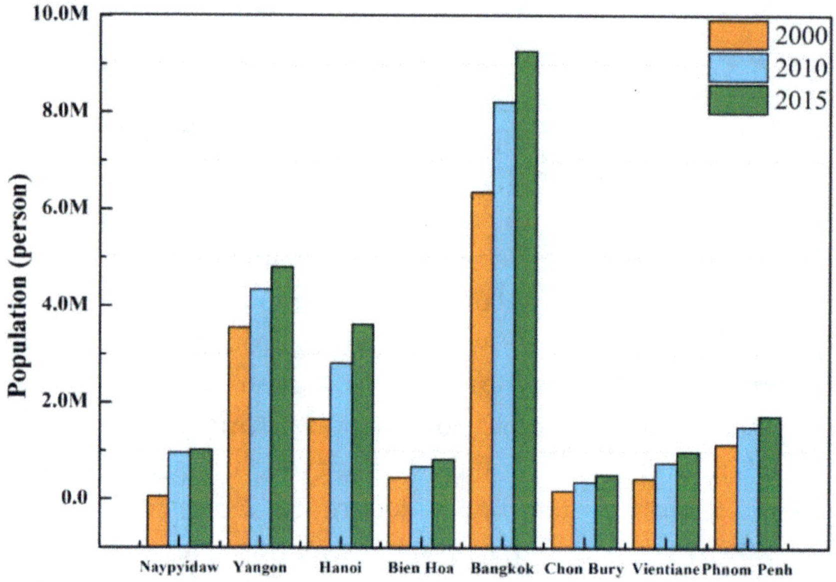

Figure 6. Map showing the UP in Naypyidaw, Yangon, Hanoi, Bien Hoa, Bangkok, Chon Bury, Vientiane, and Phnom Penh in 2000, 2010, and 2015.

Figure 7. Map showing the existing, under construction, and planned/potential railways in GMS countries in 2012.

2.2.1.3. *Infrastructure*

As seen from the development of urban areas, transport infrastructure will necessarily accelerate the expansion of urban land-use and is one of the primary driving forces of urban expansion [70]. Chen and Xia [71] also reported that a cross-regional high-speed rail network had greatly advanced China's urban development. In this study, we therefore presented a qualitative analysis of the impact of railways in the Indochinese Peninsula on urbanization during the period from 2000 to 2015. **Figure 7** shows existing, under construction, and planned/potential railways in countries of the Indochinese Peninsula in 2012. These data were obtained from the GMS Core Environment Program of ADB and were developed based on the International Vector Data and ADB maps [72]. These data provided the state of the railways in the Indochinese Peninsula around 2012 for academic research.

2.2.1.4. *Foreign economic and trade relations*

According to the econometric analysis by Huff and Angeles [73], in some Southeast Asian countries, the measures of globalization are more predictive of urbanization than domestic factors. Increasingly, the five countries in the Indochinese Peninsula are linked with the global economy through both trade and foreign direct investment (FDI) [74], and their increased outward orientation toward regional and global markets was regarded as a key contributing factor to the rapid growth during the 2000s [75]. To present a comprehensive analysis of the driving forces for urban expansion in this region, the FDI inward data and Total Merchandise Exports (TME) data were used in this chapter; they were designed to investigate the impact of foreign economic and trade relations on the region's urbanization. These two data sets were obtained from the GMS Statistics data set on the ADB Website [66], and the source of their data was the UN Conference on Trade and Development (UNCTD) [76]. **Figure 8** shows the FDI inward (a) and TME (b) for the five countries in the region.

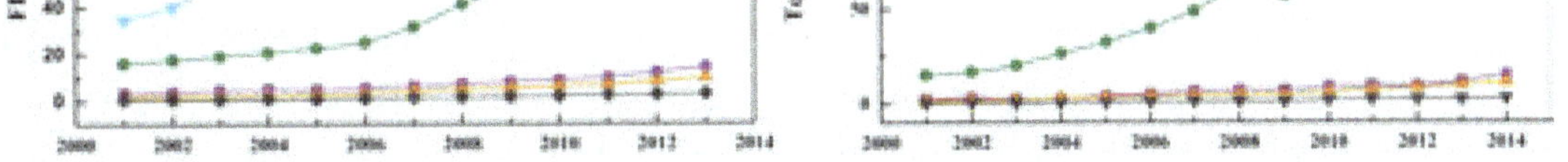

Figure 8. Map showing the FDI inward to Myanmar, Vietnam, Cambodia, Thailand, and Laos for the period from 2001 to 2013 **(a)** and TME of the five countries for the period from 2001 to 2014 **(b)**.

In addition, as a neighboring country, China has played an important role in the economic development for the five countries of the Indochinese Peninsula; in GMS, according to Poncet [77], there has been a high degree of trade linkage between China's Yunnan Province and its neighboring countries (Laos, Myanmar, and Vietnam) based on the development of a gravity model of trade. In recent years, China's outward investments in the Association of Southeast Asian Nations (ASEAN) have increased in spite of an overall global decline in FDI because of the 2008 financial crisis [78]. For this reason, China is to believed to have immense influence on the economic development in the Indochinese Peninsula. Therefore, a statistical analysis of UP (% of total), FDI from China to the five countries, and foreign trade with China was also performed in this study. FDI from China to the five countries for the period from 2003 to 2013 was used in this chapter; the data were extracted from China's Outward FDI of 2010 and 2013 [79, 80]. Data for gross imports and exports (GIE) with China for the period from 2000 to 2010 were obtained from World Integrated Trade Solution (WITS) because data from 2011 to 2015 were not obtained in China. **Figure 9** shows the FDI flowing to the five countries from China (a) for the period from 2003 to 2013 and GIE between China and the five countries (b) for the period from 2000 to 2010.

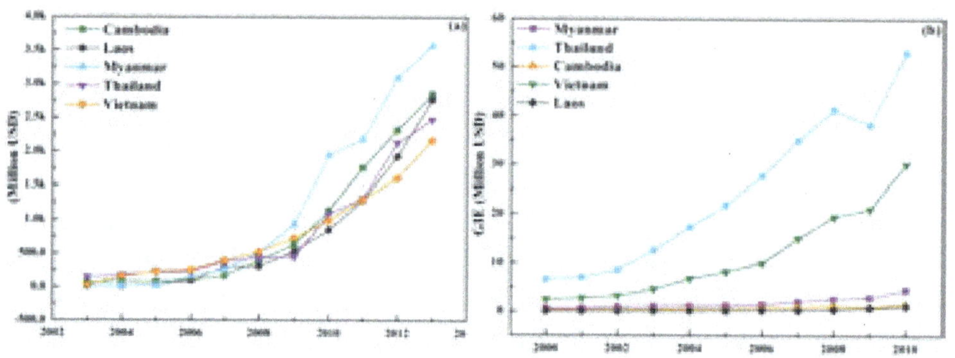

Figure 9. Map showing the FDI from China to Myanmar, Vietnam, Cambodia, Thailand, and Laos for the period from 2003 to 2013 **(a)** and GIE between China and those countries for the period from 2000 to 2010 **(b)**.

3. Result and analysis

3.1. Analysis of urban expansion in the Indochinese Peninsula

For clear information about urban expansion at the national level, the spatial data on built-up areas for the East Asian region for the period from 2000 to 2010 developed by WB were used first (data source: Platform for Urban Management and Analysis (PUMA) of WB [4]). **Figure 10** shows the built-up area in the countries of the Indochinese Peninsula in 2000 and 2010. Generally, as shown in the figure, regionwide from 2000 to 2010, the built-up area increased approximately 1963.2 km^2, expanding from 11,022.21 to 12,985 km^2. In addition, Thailand and Vietnam had much larger urban land areas in both 2000 and 2010 than the other three countries. **Table 7** shows the expansion area, expansion rate, and annual change rate for urban sprawl of the five countries for the period from 2000 to 2010. The expansion rate shows a clear heterogeneity in the region and that Thailand and Vietnam's expansion rates were higher than those of the other countries.

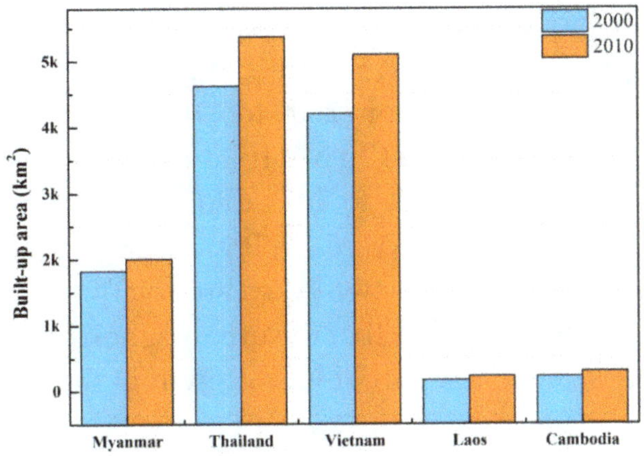

Figure 10. Built-up area (km^2) in countries of the Indochinese Peninsula in 2000 and 2010.

Item	Myanmar	Thailand	Vietnam	Laos	Cambodia
Expansion area (km^2)	182.54	749.88	897.63	60.59	72.58
Expansion rate (km^2/year)	20.28	83.32	99.74	6.73	8.06
Annual change rate (%)	1.11	1.80	2.37	4.15	3.69

Table 7. Expansion area, rate, and annual change rate for urban sprawl in countries of the Indochinese Peninsula for the period from 2000 to 2010.

Table 8 shows the built-up areas of Bangkok, Chonburi, Yangon, Naypyidaw, Hanoi, Bien Hoa, Phnom Penh, and Vientiane in 2000, 2010, and 2015 (Vientiane in 2000 and 2010), and **Table 9** shows the expansion rate and annual change rate for urban sprawl in those eight cities for the period from 2000 to 2015 (Vientiane for the period from 2000 to 2010). As shown in the tables, the built-up area of the cities in the Indochinese Peninsula, except for Phnom Penh and Vientiane, increased quickly with increased urbanization from the period from 2000 to 2015. In 2015, the built-up area of Bangkok reached 1211.55 km^2, increasing 397.19 km^2 compared to the year 2000, which shows that Bangkok has experienced a rapid urban expansion in terms of space during the past 15 years. The high annual change rate also indicates Bangkok's rapid urbanization from 2000 to 2015, especially in the past 5 years. Chonburi, Thailand, also experienced rapid urban development in the past 15 years, and the built-up area of Chonburi was approximately 94.51 km^2 in 2000 but reached 466.56 km^2 in 2015, which appears approximately five times larger than in 2000; moreover, Chonbur's annual change rate is 37.7%, the largest among the cities studied, from 2010 to 2015, which indicates a much more rapid urban development than Bangkok and the other cities in that period. In Myanmar, the built-up areas of Yangon in 2000, 2010, and 2015 are all larger than those of Naypyidaw, which indicates a high urbanization level in Yangon. Nevertheless, the expansion rate and the annual change rate of Naypyidaw are much larger than those of Yangon during the period because Myanmar's capital was moved to Naypyidaw from Yangon in 2006, thus greatly promoted urban development in the latter city. Hanoi, the capital of Vietnam, has the biggest built-up areas among the eight cities. The built-up areas of Hanoi expanded immensely in the past 15 years, increasing from 284.4 to 1164.1 km^2. The annual change rate in Hanoi from 2010 to 2015 is 36.6%, whereas it was 4.5% in the period from 2000 to 2010, which shows that Hanoi has experienced more rapid urban development in the past 5 years. Bien Hoa, as an industrial city in Vietnam, also rapidly expanded its built-up area from 2000 to 2015, with its area increasing from 58.38 to 121.54 km^2. Bien Hoa's proximity to Ho Chi Minh City and its convenient transportation are considered two important contributions to its rapid urban development. Similar to the situation in the national level, the built-up area in Phnom Penh and Vientiane are both small and the urban development levels of the two cities are relatively low. The lower expansion rate and the annual change rate also prove that the urbanization of Phnom Penh and Vientiane is slower than in other cities, and there is room for growth in their urban development.

City	2000	2010	2015
Bangkok	724.36	795.50	1121.55
Chonburi	94.51	161.60	466.56
Yangon	419.00	440.00	710.50
Naypyidaw	8.88	34.48	71.57
Hanoi	284.40	411.00	1,164.10
Bien Hoa	58.38	84.88	121.54
Phnom Penh	23.53	24.57	25.29
Vientiane	42.34	44.33	–

Table 8. Built-up area (km^2) of the eight cities selected in this study for the years 2000, 2010, and 2015.

City	2000–2010		2010–2015	
	Expansion rate (km^2/year)	Annual change rate (%)	Expansion rate (km^2/year)	Annual change rate (%)
Bangkok	7.114	1.8	65.21	8.2
Chonburi	6.709	7.1	60.992	37.7
Yangon	2.1	0.5	54.1	12.3
Naypyidaw	2.56	28.8	7.418	21.5
Hanoi	12.66	4.5	150.62	36.6
Bien Hoa	2.65	4.5	7.332	8.6
Phnom Penh	0.104	0.4	0.144	0.6
Vientiane	0.199	0.5	–	–

Table 9. Expansion rate and annual change rate for urban sprawl in the eight cities for the period from 2000 to 2015.

Table 10 shows the increased and annual change rate of population growth in cities with populations of more than 300,000 in the Indochinese Peninsula. In general, the total increased population in Thailand is approximately 3,419,230 persons for the period from 2000 to 2010

Cities	Country	Population growth from 2000 to 2010	Annual change rate from 2000 to 2010	Population growth from 2010 to 2015	Annual change rate from 2010 to 2015
Bangkok		1852.88	2.9	1056.46	1.3
Chonburi		184.14	9.9	147.49	4.0
Hat Yai		79.16	4.2	48.97	1.8
Lampang		131.20	8.7	100.55	3.6
Nakon Patchasima	**Thailand**	98.40	4.8	62.90	2.1
Nonthaburi		74.57	2.5	41.40	1.1
Rayong		121.53	11.2	101.92	4.4
Samut Prakan		703.96	18.1	721.48	6.6
Udon Thani		173.39	7.7	127.41	3.2
Bien Hao		228.45	5.0	147.83	2.2
Can Tho		412.09	9.4	324.05	3.8
Da Nang		237.35	4.2	146.88	1.8
Hanoi		1151.28	6.9	818.47	2.9
Haiphong	**Vietnam**	290.34	4.8	186.50	2.1
Hue		71.47	3.0	40.97	1.3
Nha Trang		32.09	1.2	15.61	0.5
Ho Chi Minh City		1800.00	4.1	1108.41	1.8
Vung Tau		89.59	4.4	56.01	1.9
Bago		136.10	4.7	90.13	2.1
Mandalay		222.81	2.7	133.67	1.3
Mawlamyine	**Myanmar**	93.82	2.8	56.43	1.3
Monywa		140.95	5.9	98.89	2.6
Naypyidaw		908.04	162	65.68	0.7
Yangon		789.40	2.5	459.79	1.2
Vientiane	**Laos**	318.99	7.2	235.45	3.1
Phnom Penh	**Cambodia**	361.01	2.2	221.57	1.1

Table 10. Increased (1000 persons) and annual change rate (%) for population growth in cities of the Indochinese Peninsula with populations of more than 300,000.

and 2,408,580 persons from 2010 to 2015. The mean of the annual change rate is approximately 7.78% from 2000 to 2010 and approximately 3.12% from 2010 to 2015, which shows that the UP increased much more slowly in the past 5 years than in the past. It is particularly necessary to note that Samut Prakan, Rayong, and Lampang's UPs grew faster than in the other cities from 2000 to 2010, and this is also the case during the most recent 5 years. The high annual change rate indicates that the three cities have higher urbanization levels than other cities. For cities with populations of more than 300,000 in Vietnam, the UP is also increasing rapidly, with the number growing by 4,312,660 for the period from 2000 to 2010 and 2,844,730 from 2010 to 2015. The mean of the annual change rate is approximately 4.78% and 2.03% for the period from 2000 to 2010 and from 2010 to 2015, respectively, and UP growth also slowed in recent years. In Thailand, no city showed an obvious population increase for the past 15 years, except for Can Tho. There are six cities with populations of more than 300,000 in Myanmar, and the increased population in those cities is 2,291,120 for the period from 2000 to 2010 and 904,590 for the period from 2010 to 2015. The mean of the annual change rate is 30.1% for the period from 2000 to 2010 but only 1.53% from 2010 to 2015. It should be noted that the annual change rate in Naypyidaw for the period from 2000 to 2010 is far greater than in other cities, which proves once again that moving the capital significantly contributed to population growth in Naypyidaw. Unlike the situation of urban expansion, the population of Vientiane, Laos, grew rapidly during the past 15 years; this growth was more obvious in the previous decade. Phnom Penh's population only increased by approximately 582,580 persons over the past 15 years and its population growth rate remains lower compared to other cities during that period.

3.2. Analysis of driving forces for urban expansion in the Indochinese Peninsula

3.2.1. Geographical elements

Geographical location, including absolute and relative location, plays an important role in the formation and development of the city, and the correlation between urban growth and geographical location is primarily reflected in the interaction between urban and geographic elements. A city's location is the characteristic of the combination of the city, nature, politics, and economics in space, and a favorable geographic location will promote urban development. In addition, the urban area's geographical location can also decide the specificity of the city's function and size. The urban development in the countries of the Indochinese Peninsula is generally affected by their relative geographical location, and cities with populations of more than 500,000 are primarily distributed around the coastal areas of the Peninsula (see **Figure 1**). Ho Chi Minh City, the largest city in Vietnam, is one of the world's largest seaports, and its urban development is primarily credited to its favorable geographical location, which is close to the rivers. Haiphong, Vietnam's urban development is largely influenced by the international maritime services. In addition, a favorable geographical location is also considered as one of the key factors in the urban development of Yangon, Myanmar. In any event, geographical location has determined the development of cities in the Indochinese Peninsula, at least to some extent.

3.2.2. Transport infrastructure elements

There is a complicated relationship between urban development and the transport infrastructure, and urban development creates many advanced vehicles to improve the urbanization process. The influence of the transport infrastructure on urban development primarily emanates from two important aspects. First, we consider metropolitan transportation and exterior traffic conditions. Very convenient transportation conditions can optimize the industrial layout of the city, and its changes can directly affect the city's structure and industrial layout. Second, convenient transportation conditions also have a substantial impact on economic development, thus accelerating urbanization. Moreover, the direction of emigration is decided by the transport infrastructure, and convenient transportation conditions provide opportunities for labor-force exchanges. In the cities of the Indochinese Peninsula, in addition to the influence of seaway transportation, regional railways' transportation conditions determine urban development. **Figure** 7 shows that the big cities in the region all follow convenient railway transportation. This phenomenon can be better illustrated by the urban development differentials among the five countries. As mentioned above, the level of urban development in the five countries reveals heterogeneity, and the urbanization process for the period from 2000 to 2015 varies by country. Thailand, Vietnam, and Myanmar, which have more developed railway networks, show much stronger capabilities in their urban development than do Laos and Cambodia. To improve the railways, basic facilities construction is essential for urbanization in Laos and Cambodia. Furthermore, with the development of high-speed rail, the cities in the Indochinese Peninsula will obtain a new development opportunity.

3.2.3. Economic growth elements

Most empirical studies report that economic growth promoted the increase of both the built-up areas and UP growth, and there is a strong correlation between economic growth and urbanization [81, 82]. **Table 11** shows the summary statistical results from the linear regression model examining the relationship between urbanization and GDP in the countries of the Indochinese Peninsula during the period from 2001 to 2014. In general, we learn that there is an obvious correlation between UP (% of total) and GDP in each country, with an average value of 0.981. Additionally, there are no significant differences in the R values among the countries. Furthermore, the functions of the linear regression model indicate that UP grows with the increase of GDP, demonstrating the most direct influence on UP growth by economic growth in the cities.

Country	**Dependent variable (y)**	**Independent variable (x)**	R	**Function**[a]
Myanmar	UP (% of total)	GDP (billion USD)	0.982	$y = 0.09x + 27.23$
Thailand			0.983	$y = 0.05x + 27.67$
Vietnam			0.986	$y = 0.05x + 24.23$
Laos			0.970	$y = 1.42x + 22.67$
Cambodia			0.985	$y = 0.13x + 18.3$

[a]The functions were valid because they all passed the F-test, and all of the regression coefficients passed the t-test (at the level of 0.05).

Table 11. Relationship between urbanization in the countries of the Indochinese Peninsula and GDP during the period from 2001 to 2014 ($n = 14$).

3.2.4. Foreign economic and trade relations

Table 12 shows the summary statistical results from the linear regression model examining the relationship between UP (% of total) and FDI in the five countries. High *R* values appear in the five countries, thus indicating that the two items are highly correlated. The functions also illustrate that increased FDI could drive urban development in the countries of the Indochinese Peninsula. Much like the situation in the previous model, high *R* values between UP (% of total) and TME are obtained in all countries (**Table 13**), denoting that there is a positive correlation between them. From the results of the functions, we further find that boosting the export value can promote urban development in the five countries. Moreover, from the summary statistical results for both scenarios, we learn that the influences from FDI and TME on UP growth are at almost the same level because there is little difference in the values of the Pearson correlation coefficient. FDI and foreign trade have played an almost equally important role in the urban development of the five countries for the period from 2000 to 2015. Furthermore, there is little difference among the various nations in the *R* values in both scenarios, which indicates that the continued outward orientation towards the global market and the absorption of foreign investment are considered key contributing factors to the rapid urbanization of each country of the Indochinese Peninsula during the studied period.

Country	**Dependent variable (y)**	**Independent variable (x)**	*R*	**Function**[a]
Myanmar	UP (% of total)	FDI (billion USD)	0.964	$y = 0.54x + 26.17$
Thailand			0.986	$y = 0.10x + 30.40$
Vietnam			0.968	$y = 0.10x + 24.51$
Laos			0.961	$y = 5.48x + 22.26$
Cambodia			0.979	$y = 0.20x + 18.58$

[a]The functions were valid because they all passed the *F*-test, and all of the regression coefficients passed the *t*-test (at the level of 0.05).

Table 12. Relationship between urbanization in the countries of the Indochinese Peninsula and FDI during the period from 2001 to 2013 ($n = 13$).

Country	**Dependent variable (y)**	**Independent variable (x)**	*R*	**Function**[a]
Myanmar	UP (% of total)	TME (billion USD)	0.962	$y = 0.52x + 26.82$
Thailand			0.979	$y = 0.08x + 27.92$
Vietnam			0.964	$y = 0.06x + 25.31$
Laos			0.964	$y = 5.38x + 23.75$
Cambodia			0.976	$y = 0.20x + 18.58$

[a]The functions were valid because they all passed the *F*-test, and all of the regression coefficients passed the *t*-test (at the level of 0.05).

Table 13. Relationship between urbanization in the countries of the Indochinese Peninsula and TME during the period from 2001 to 2014 ($n = 14$).

3.2.5. Trade and investment linkage with China

Tables 14 and **15** show both the results of the linear regression model examining the relationship between urbanization and FDI from China during the period from 2003 to 2013 and the results examining the relationship between urbanization and GIE with China during the period from 2000 to 2010 in the countries of the Indochinese Peninsula. As shown in **Table 14**, the UP is highly correlated with FDI from China in the five countries, and the high *R* values indicate that urbanization is sensitive to increased investment from China. In addition, the functions derived from the analytical models demonstrate that investment from China is a key driving force for UP growth in the region. According to the analytical results of the relationship between UP (% of total) and GIE with China, we can obtain a similar situation with the aforementioned scenario, and the two items are highly correlated with each other in each country of the Indochinese Peninsula. Moreover, the derived functions show both that trade with China has positive effects on UP growth in the five countries and that increasing bilateral trade will be beneficial to facilitate urbanization in the region. Briefly, trade and investment with China have a substantial effect on the urbanization process in the countries of the Indochinese Peninsula during the study period, and improving the economic cooperation between China and the five countries will contribute to the region's urbanization.

Country	Dependent variable (y)	Independent variable (x)	R	Function[a]
Myanmar	UP (% of total)	FDI from China (million USD)	0.946	$y = 0.001x + 29.07$
Thailand			0.950	$y = 0.005x + 37.60$
Vietnam			0.952	$y = 0.003x + 26.96$
Laos			0.892	$y = 0.004x + 28.08$
Cambodia			0.939	$y = 4.2x + 19.21$

[a]The functions were valid because they all passed the *F*-test, and all of the regression coefficients passed the *t*-test (at the level of 0.05).

Table 14. Relationship between urbanization in the countries of the Indochinese Peninsula and FDI from China during the period from 2003 to 2013 ($n = 11$).

Country	Dependent variable (y)	Independent variable (x)	R	Function[a]
Myanmar	UP (% of total)	GIE with China (million USD)	0.934	$y = 1.17x + 27.00$
Thailand			0.979	$y = 0.26x + 31.19$
Vietnam			0.950	$y = 0.21x + 24.97$
Laos			0.854	$y = 9.54x + 24.64$
Cambodia			0.961	$y = 0.95x + 18.55$

[a]The functions were valid because they all passed the *F*-test, and all of the regression coefficients passed the *t*-test (at the level of 0.05).

Table 15. Relationship between urbanization in the countries of the Indochinese Peninsula and GIE with China during the period from 2000 to 2010 ($n = 11$).

3.3. Sustainable urbanization in the China-Indochinese Peninsula Economic Corridor

The coordinated development between urban area expansion and UP growth is one important aspect of sustainable urbanization [83]. In this chapter, we use Spearman rank correlation to test the association between the annual change rate of urban expansion and population growth in the Indochinese Peninsula for the period from 2000 to 2015. According to the result of the Spearman model, we determine that the correlation coefficient between the two items is 0.807 (Sig.=0.015) for the period from 2000 to 2010 and 0.536 (Sig.=0.215) for the period from 2010 to 2015. It is obvious that the annual change rates for urban expansion and population growth are highly correlated with each other for the period from 2000 to 2010 among the cities in the Indochinese Peninsula; however, for the period from 2010 to 2015, the statistical analysis result does not pass significance testing, which indicates that there has been disequilibrium developing between urban expansion and population growth in the region since 2010. Thus, in the future, more attention should be paid to sustainable urbanization in the Indochinese Peninsula.

Based on the analysis of driving forces for urban expansion, we conclude that, in addition to socioeconomic factors, FDI and international trade have a noticeable correlation with urban development in the countries of the Indochinese Peninsula. As we have discussed above, railway infrastructure construction is significant to urbanization in countries of the Indochinese Peninsula. However, some previous studies reported that the lack of financial resources was a serious obstacle to infrastructure development in Southeast Asia, and transportation infrastructure in the region could not meet the need for urban development because of fiscal pressures [84]. With assistance from the ADB and the WB, Asian countries receive approximately 20 billion USD in annual fiscal support but still cannot maintain basic transport infrastructure investments in the railways, airports, seaports, roads, and communication facilities needed for urban development [85]. Therefore, new investment has become a key factor, especially in Cambodia and Laos. The Asian Infrastructure Investment Bank (AIIB), proposed by President Jinping Xi of China on October 2, 2013, primarily aims to provide financing for infrastructure projects in the Asia Pacific region [86]. Increased investment by the AIIB will be very helpful for the construction of urban infrastructure and the promotion of urban development. Furthermore, the trade cooperation between the five countries and China is an important contributing factor to the forces driving urbanization at least according to the statistical analysis. With the benefit of the investment and trade opportunity provided by B&R, the China-Indochinese Peninsula Economic Corridor will embrace new and increased opportunities for urban development.

4. Conclusion

This chapter presents a review of the urbanization in countries of the Indochinese Peninsula using advanced remote sensing technology. It also analyzes the driving forces for urban expansion. Our conclusions are as follows:

1. The urbanization progress increased rapidly in the Indochinese Peninsula region both in terms of urban areas and UP; however, the level of urban development in countries of the

Indochinese Peninsula appears to represent a spatial heterogeneity; Thailand and Vietnam have expanded rapidly in urban land compared to the other countries in the study period, whereas Laos remains at a low level of development. Overall, the urbanization of the Mekong countries remains broad.

2. In addition to socioeconomic factors, FDI and international trade also have a noticeable correlation with urban development in the Indochinese Peninsula. Foreign investment plays a significant role in regional urbanization.

3. Investment from China increased quickly in the past 5 years and has close relationship with regional urbanization rate. China, as a neighbor to the Mekong countries, will play an increasingly important role in their urbanization.

4. The China-Indochinese Peninsula Economic Corridor will witness a more rapid urbanization progress in the next decade primarily because of the launch of the Silk Road Economic Belt and the 21st Century Maritime Silk Road and the AIIB.

5. The primary advantages of the manuscript include the following:

    - Its focus on the areas of the Indochinese Peninsula in which the most rapid urbanization is occurring.
    - It adoption of the latest and most precise data set.
    - Its integrated analysis that employed multiple principles and multiple-level data.

Author details

Dong Jiang, Jingying Fu* and Gang Lin

*Address all correspondence to: fujy@igsnrr.ac.cn

State Key Laboratory of Resources and Environmental Information System, Institute of Geographical Sciences and Natural Resources Research, Chinese Academy of Sciences, Beijing, China

References

[1] UN. *World Urbanization Prospects: The 2014 Revision*. 2015. New York: United Nations Publications.

[2] CESA. *100 Cities Project*. 2010. ASU.

[3] SEDAC. *Global Rural-Urban Mapping Project*. 2012. NASA.

[4] WB. *Platform for Urban Management and Analysis*. 2015. https://puma.worldbank.org/ [cited 2015/29/9].

[5] Bagan, H. and Y. Yamagata. *Land-cover change analysis in 50 global cities by using a combination of Landsat data and analysis of grid cells*. Environmental Research Letters, 2014, 9(6), p, 064015.

[6] Schneider, A., M.A. Friedl, and D. Potere. *A new map of global urban extent from MODIS satellite data*. Environmental Research Letters, 2009, 4(4), p, 044003.

[7] Asian Development Bank. *Greater Mekong Subregion Economic Cooperation Program*. 2015. http://www.adb.org/sites/default/files/publication/29387/gms-ecp-overview-2015.pdf [cited 2015/11/12].

[8] ADB. *Overview of the Greater Mekong Subregion*. 2012. http://www.adb.org/countries/gms/overview [cited 2016/16/1].

[9] ADB. *Greater Mekong Subregion: Twenty Years of Partnership*. 2012. http://www10.iadb.org/intal/intalcdi/PE/2012/11315.pdf [cited 2016/16/1].

[10] Yang, M., L. Yang, and Z. Hu. *Pattern evolution and challenges of trades between China and Indo-China Peninsula countries under the background of "One Belt and One Road"* (in Chinese with English abstract). Tropical Geography, 2015, 35(5), p. 655–663.

[11] Jat, M.K., P.K. Garg, and D. Khare. *Monitoring and modelling of urban sprawl using remote sensing and GIS techniques*. International Journal of Applied Earth Observation and Geoinformation, 2008, 10(1), p. 26–43.

[12] Morawitz, D.F., et al. *Using NDVI to assess vegetative land cover change in central Puget Sound*. Environmental Monitoring and Assessment, 2006, 114(1–3), p. 85–106.

[13] Nagendra, H. and G. Utkarsh, *Landscape ecological planning through a multi-scale characterization of pattern: Studies in the Western Ghats, South India*. Environmental Monitoring and Assessment, 2003, 87(3), p. 215–233.

[14] Yuan, F., et al. *Land cover classification and change analysis of the Twin Cities (Minnesota) Metropolitan Area by multitemporal Landsat remote sensing*. Remote Sensing of Environment, 2005, 98(2–3), p. 317–328.

[15] Taubenböck, H., et al. *Urbanization in India—Spatiotemporal analysis using remote sensing data*. Computers, Environment and Urban Systems, 2009, 33(3), p. 179–188.

[16] Yeh, A.G.O. and X. Li. *Measurement and monitoring of urban sprawl in a rapidly growing region using entropy*. Photogrammetric Engineering and Remote Sensing, 2001, 67(1), p. 83–90.

[17] Yang, X.J. and C.P. Lo. *Modelling urban growth and landscape changes in the Atlanta metropolitan area*. International Journal of Geographical Information Science, 2003, 17(5), p. 463–488.

[18] Haack, B.N. and A. Rafter. *Urban growth analysis and modeling in the Kathmandu Valley, Nepal*. Habitat International, 2006, 30(4), p. 1056–1065.

[19] Dewan, A.M. and Y. Yamaguchi. *Using remote sensing and GIS to detect and monitor land use and land cover change in Dhaka Metropolitan of Bangladesh during 1960–2005*. Environmental Monitoring and Assessment, 2009, 150(1–4), p. 237–249.

[20] Kong, I.-H., G.-H. Baek, and D.-K. Lee. *Land cover change and forest fragmentation analysis for Naypyidaw, Myanmar*. Journal of Environmental Impact Assessment, 2013, 22(2), p. 147–156.

[21] Ourng, C. and D.S. Rodrigues. Urban growth pattern identification: A case study in Siem Reap, Cambodia. In *PLURIS 2012-5 Congresso Luso-Brasileiro para o Planejamento Urbano, Regional, Integrado e Sustentável*. 2012. Universidade de Brasília, Faculdade de Arquitetura e Urbanismo, Brasilia, Brazil.

[22] Okamoto, K., A. Sharifi, and Y. Chiba. The impact of urbanization on land use and the changing role of forests in Vientiane. S. Yokoyama et al. (eds.), *In Integrated Studies of Social and Natural Environmental Transition in Laos*. 2014. Springer Japan. p. 29–38.

[23] Kimijiama, S. and M. Nagai. Study for urbanization corresponding to socio-economic activities in Savannaket, Laos using satellite remote sensing. In *IOP Conference Series: Earth and Environmental Science*. 2014. IOP Publishing, Kuala Lumpur, Malaysia.

[24] Jing, G. *The Theoretical Issues of Landscape Ecology*. 1991. Beijing: China Forestry Publishing (in Chinese).

[25] Shi, P. *The Methods and Practices in the Studies of LUCC*. 2000. Beijing: Science Press (in Chinese).

[26] Zhao, Y. *The Principle and Method for the Analysis of Remote Sensing Applications*. 2003. Beijing: Science Press. p. 1–6. (in Chinese).

[27] Kurkure, U., et al. *A supervised classification-based method for coronary calcium detection in non-contrast CT*. International Journal of Cardiovascular Imaging, 2010, 26(7), p. 817–828.

[28] Jiang, P., Q.S. Dou, and X.Y. Hu. *A supervised method for retinal image vessel segmentation by embedded learning and classification*. Journal of Intelligent & Fuzzy Systems, 2015, 29(5), p. 2305–2315.

[29] Li, X., et al. *Supervised transfer kernel sparse coding for image classification*. Pattern Recognition Letters, 2015, 68, p. 27–33.

[30] Deng, J., et al. *Feature extraction and classification of Tilletia diseases based on image recognition*. Transactions of the CSAE, 2012, 28(3), p. 172–176 (in Chinese with English abstract).

[31] Wang, X., Q. Yin, and P. Guo. *Text-independent speaker identification using Fisher discrimination dictionary learning method*. Proceedings of the 2012 Eighth International Conference on Computational Intelligence and Security (Cis 2012), 2012, p. 435–438.

[32] Kucewicz, J.C., L.Y. Huang, and K.W. Beach. *Plethysmographic arterial waveform strain discrimination by Fisher's method.* Ultrasound in Medicine and Biology, 2004, 30(6), p. 773–782.

[33] Yang, W., T. Fang, and G. Xu. *Semi-supervised learning remote sensing image classification based on naive Bayesian.* Computer Engineering, 2010, 36(20), p. 167–169 (in Chinese with English abstract).

[34] Xu, X., et al. *Classification application of Quickbird imagery to obtain crop planting area.* Remote Sensing Technology and Application, 2008, 23(1), p. 17–25 (in Chinese with English abstract).

[35] Wu, F., et al. *Combining the decision tree and supervised, unsupervised technique to classify the satellite images.* Journal of Salt Lake Research, 2005, 4(13), p. 9–13 (in Chinese with English abstract).

[36] Lee, T.W. and M.S. Lewicki. *Unsupervised image classification, segmentation, and enhancement using ICA mixture models.* IEEE Transactions on Image Process, 2002, 11(3), p. 270–279.

[37] Cheong, C., G. Bowman, and T.D. Han. *Unsupervised clustering approaches to color classification for color-based image code recognition.* Applied Optics, 2008, 47(13), p. 2326–2345.

[38] Aschkenasy, S.V., et al. *Unsupervised image classification of medical ultrasound data by multiresolution elastic registration.* Ultrasound in Medicine & Biology, 2006, 32(7), p. 1047–1054.

[39] Zha, Y. and S. Ni. *An effective approach to automatically extract urban Land-use from TM imagery.* Journal of Remote Sensing, 2003, 7(1), p. 37–40 (in Chinese with English abstract).

[40] Xue, H. and S. Ni. *Methodology of retrieving residential area from remotely sensed image.* Research of Soil and Water Conservation, 2006, 13(5), p. 319–321 (in Chinese with English abstract).

[41] Pan, W. and Zhang, C. *A study on urban heat island effect in Quanzhou City during its urbanization period. Remote Sensing for Land and Resources,* 2006, 4(70), p. 50–55 (in Chinese with English abstract).

[42] Stevens, K.A. *The visual interpretation of surface contours.* Artificial Intelligence, 1981, 17(1), p. 47–73.

[43] Cai, L. *Talk about visual interpretation of remote sensing image and its influence factors in urban planning.* Chinese and Overseas Architecture, 2013(6), p. 82–84 (in Chinese with English abstract).

[44] Ma, A. *Basic theory and methods for visual interpretation by remote sensing.* Remote Sensing Information, 1987. 3, p. 26–29 (in Chinese).

[45] Jiang, F. and G. Zhang. *Several notable issues on visual interpretation of remote sensing image*. Journal of Changchun Institute of Technology, 2002, 3(3), p. 49–50 (in Chinese with English abstract).

[46] Jiang, D., et al. *A simple semi-automatic approach for land cover classification from multispectral remote sensing imagery*. PLoS ONE, 2012, 7(9), p. e45889.

[47] Wang, Y., et al. *Effective key parameter determination for an automatic approach to land cover classification based on multispectral remote sensing imagery*. PLoS ONE, 2013, 8(10), p. e75852.

[48] Yang, S. *On extraction and fractal of urban and rural residential spatial pattern in developed area*. Journal of Geographical Sciences, 2000, 55(6), p. 671–678 (in Chinese with English abstract).

[49] Meng, F., M. Liu, and X. Zhang, *Extracting urban cover and background information on ETM images*. Journal of East China Normal University, 2005, 4, p. 59–65 (in Chinese with English abstract).

[50] Dong, G., et al. *Remote sensing image classification algorithm based on Hopfield neural network*. Journal of Institute of Surveying and Mapping, 2004, 20(4), p. 267–270.

[51] Chen, X., et al. *Research on direct classification change detection for temporal and multispectral remote sensing data using neural network*. Computers Engineering and Applications, 2004, 40(28), p. 12–15 (in Chinese with English abstract).

[52] Li, S. and D. Zheng. *Application of artificial neural networks to geosciences: Review and prospect*. Advances in Earth Sciences, 2003, 18(1), p. 68–76 (in Chinese with English abstract).

[53] Irish, R.R. *Landsat 7 Science Data Users Handbook*. National Aeronautics and Space Administration Report, 2000, p. 430–415.

[54] Roy, D.P., et al. *Landsat-8: Science and product vision for terrestrial global change research*. Remote Sensing of Environment, 2014, 145, p. 154–172.

[55] Wang, L. and Y. Bu. *The application and development of the high-resolution remote sensing satellite*. Geomatics Technology and Equipment, 2007, 8(4), p. 3–5 (in Chinese).

[56] Lillesand, T., R. Kiefer, and J. Chipman. *Remote Sensing and Image Analysis*. 2000. New York: John Wiley and Sons.

[57] Tong, Q., B. Zhang, and L. Zheng. *Multi-Disciplinary Applications for Hyperspectral Remote Sensing*. 2006. Beijing: Electronic Industry Press. p. 196 (in Chinese).

[58] Epstein, J., K. Payne, and E. Kramer. *Techniques for mapping suburban sprawl*. Photogrammetric Engineering and Remote Sensing, 2002, 68(9), p. 913–918.

[59] Xu, X. and X. Min. *Quantifying spatiotemporal patterns of urban expansion in China using remote sensing data*. Cities, 2013, 35, p. 104–113.

[60] Xiao, J., et al. *Evaluating urban expansion and land use change in Shijiazhuang, China, by using GIS and remote sensing*. Landscape and Urban Planning, 2006, 75(1–2), p. 69–80.

[61] Zhang, X. *The dynamic research of urban expansion based on RS and GIS*. 2007, Tongji University, Shanghai. p. 67 (in Chinese with English abstract).

[62] Lu, X. *Dynamic monitoring of urban expansion based on RS and GIS technology—A case study of Zhengzhou City*. Journal of Henan Polytechnic University, 2008, 27(2), p. 182–187 (in Chinese with English abstract).

[63] Wang, L., et al. *China's urban expansion from 1990 to 2010 determined with satellite remote sensing*. Chinese Science Bulletin, 2012, 57(22), p. 2802–2812.

[64] Mundia, C.N. and M. Aniya. *Analysis of land use/cover changes and urban expansion of Nairobi City using remote sensing and GIS*. International Journal of Remote Sensing, 2005, 26(13), p. 2831–2849.

[65] IMF. *World Economic Outlook*. 2015. Bernan Associates, Washington, USA.

[66] ADB. *GMS Statistics. Gross domestic product in GMS countries 2015*. http://www.gms-eoc.org/gms-statistics/overview/gross-domestic-product-current-prices- [cited 2016/4/1].

[67] Stankowski, S.J. *Population density as an indirect indicator of urban and suburban land-surface modifications*. US Geological Survey Professional Paper, 1972, 800, p. 219–224.

[68] López, E., et al. *Predicting land-cover and land-use change in the urban fringe: A case in Morelia City, Mexico*. Landscape and Urban Planning, 2001, 55(4), p. 271–285.

[69] WB. *World Development Indicators—Urban Population (% of Total)*. 2015. http://data.worldbank.org/indicator/SP.URB.TOTL.IN.ZS [cited 2015/3/10].

[70] Ma, Y.L. and R.S. Xu. *Remote sensing monitoring and driving force analysis of urban expansion in Guangzhou City, China*. Habitat International, 2010, 34(2), p. 228–235.

[71] Gang, C.L.X. *The significance of rail transportation to the urbanization of China*. Urban Problems, 2008, 2, p. 005.

[72] ADB. *GMS Railways: Status (Existing, Under Construction, Planned/Potential)*. 2012. www.gms-eoc.org.

[73] Huff, G. and L. Angeles. *Globalization, industrialization and urbanization in pre-World War II Southeast Asia*. Explorations in Economic History, 2011, 48(1), p. 20–36.

[74] Termpittayapaisith, A. and L. Kumpa. *Dynamics of economic growth in the GMS: A retrospective view 2000–2010*. GMS 2020 Conference Proceedings, Bangkok, Thailand, 2012, p. 10–11.

[75] Kumar, U. and P. Srivastava. Growth in the Greater Mekong Subregion in 2000–2010 and future prospects. In *GMS 2020 Conference Proceedings, Bangkok, Thailand*, 2012, p. 12–36.

[76] UN. *UNCTADstat*. 2015. http://unctadstat.unctad.org/wds/ReportFolders/reportFolders.aspx?sCS_ChosenLang=en [cited 2015/26/12].

[77] Poncet, S. *Economic integration of Yunnan with the Greater Mekong Subregion.* Asian Economic Journal, 2006, 20(3), p. 303.

[78] Salidjanova, N. and US-China Economic and Security Review Commission. *Going Out: An Overview of China's Outward Foreign Direct Investment*. 2011. US-China Economic and Security Review Commission, Beijing, China.

[79] MOFCOM. *2013 Statistical Bulletin of China's Outward Foreign Direct Investment*. 2014. Ministry of Commerce, Beijing, China (in Chinese).

[80] MOFCOM. *2010 Statistical Bulletin of China's Outward Foreign Direct Investment*. 2011. Ministry of Commerce, Beijing, China (in Chinese).

[81] Wellisz, S.H. *Economic Development and Urbanization*. SAGE Publications Inc., California. 1971.

[82] Dima, B., N.C. Leitao, and S. Dima. *Urbanization and Democracy in the Framework of Modernization Theory: Recent Empirical Evidences*. National Academy of Management Vul., Ukraine. 2011.

[83] Zhu, Y. and Y. Chen. *The relationship between population urbanization and urban sprawl and "Compact City"*. Modern Economic Information, 2014, 10, p. 014 (in Chinese).

[84] Dahiya, B. *Southeast Asia and sustainable urbanization.* Global Asia, 2014, 9(3), p. 84–91.

[85] Wang, L. *Roadmap guess of AIIB.* International Financial News, 2014, p. S.24.

[86] AP. *Twenty Other Countries Initiate New Asian Bank. Associated Press, Beijing*. 2014.

10

Towards Sustainable Sanitation in an Urbanising World

Philippe Reymond, Samuel Renggli and Christoph Lüthi

Abstract

Urban sanitation in low- and middle-income countries is at an inflection point. It is increasingly acknowledged that conventional sewer-based sanitation cannot be the only solution for expanding urban areas. There are other objective reasons apart from the lack of capital. The lack of stable energy supplies, of spare parts and of human resources for reliable operation, and the increasing water scarcity are factors that seriously limit the expansion of centralised systems. This chapter argues that a new paradigm for urban sanitation is possible, if the heterogeneity within developing cities is reflected in the implementation of different sanitation systems, adapted to each urban context and integrated under one institutional roof. This new paradigm entails: (1) innovative management arrangements; (2) increased participation and the integration of individual, community and private sector initiatives; (3) thinking at scale to open new opportunities; (4) improved analysis of the situation and awareness raising. Moving beyond conventional approaches towards sustainable urbanisation needs to follow both a top-down and a bottom-up approach, with proper incentives and a variety of sanitation systems which, in a future perspective, will become part of the 'urban ecosystem'.

Keywords: urban sanitation, sanitation planning, decentralised sanitation, wastewater management, faecal sludge management

1. Introduction

The world is experiencing unprecedented urban population growth rates, and most of this growth is projected to take place in cities of low- and middle-income countries in Africa and Asia. The United Nation's latest global population estimates, published in early 2015, project that the global population in 2100 will be 11.2 billion. It is estimated that Africa will have 39% of the world's population, almost as much as is estimated for Asia [1].

The rapidly urbanising societies of Africa, Asia and Latin America are constricted by a quadruple challenge: urban environmental degradation, global climate change with accentuating water stress, infrastructure deficits and fast expanding peri-urban areas and informal settlements. As pointed out by [2], even when urban sanitation management infrastructure is available, it often serves only a small percentage of the urban population. Small- and medium-sized towns (<500,000 inhabitants) will carry the brunt of future urbanisation in low- and middle-income countries and will have a pronounced backlog in urban sanitation infrastructure. A recent infrastructure study of the World Bank highlighted the low access to improved sanitation in urban Africa with 51% of the population relying on traditional (unimproved) latrines, 14% on improved latrines and only 25% connected to sewers or a septic tank [3].

Urban sanitation is at an inflection point. The international community and national governments are increasingly acknowledging that conventional sewer-based sanitation cannot be the solution for all the different urban areas. Apart from the lack of capital, there are other good objective reasons why conventional urban water management does not offer the only solution for the rapidly growing cities in Asia and Africa: the lack of stable energy supplies, of spare parts and of human resources for reliable operation are factors that limit the expansion of centralised systems. In an increasing number of cities, water scarcity is also becoming an important bottleneck. As a special case, the improvement of sanitation conditions in informal settlements in low- and middle-income countries has proven difficult due to disabling institutional environments, as well as the lack of secure tenure and of the rule of law, often preventing the development of innovative management schemes. Today, a majority of urban citizens rely on on-site systems, such as septic tanks, pit latrines or cesspits. With sewer-based systems out of reach for a large part of the global population, there is an urgent need to develop more cost-effective and resource-efficient systems that can deliver the desired water services necessary for public health, protection against flooding, and the preservation of natural resources. In this chapter, we present the main reasons for environmental sanitation deficits and lay out arguments for a holistic sectoral approach that is inclusive and that incorporates innovative management arrangements for growing urban areas.

2. Different urban contexts—different sanitation challenges

Cities are not homogeneous, especially in low- and middle-income countries. Lüthi et al. [4] identified four typical urban contexts: (1) inner-city middle and high-income settlements, (2) planned urban development areas, (3) informal settlements and (4) peri-urban interface (see **Figure 1**). Contrasts between these contexts can be striking, that is skyscrapers and slum pockets in India or favelas next to villas with swimming pools in Brazil. It is currently common to find modern city centres next to informal neighbourhoods that lack the most basic services. The tendency in current urbanisation is an increase in segregation, with a densification and sprawling of informal settlements on the one side, and the rapid extension of medium density planned urban development areas, that is large fully fledged new neighbourhoods or gated communities for the middle and upper classes, on the other.

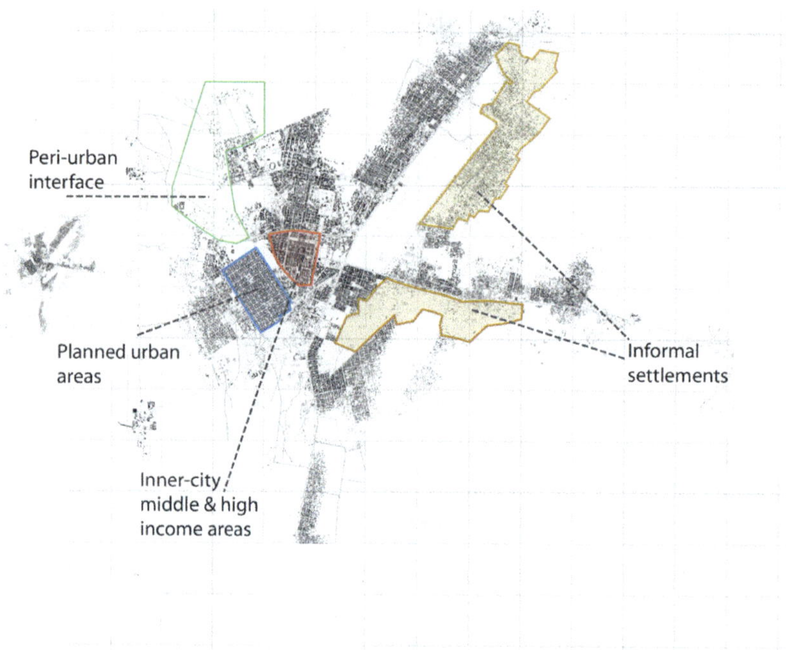

Figure 1. Main settlement contexts that need to be addressed in urban sanitation in low- and middle-income countries (Source: [4], p. 79).

Inner-city middle- and high-income settlements, as well as planned urban development areas, are usually characterised as having conventional centralised water supply and sewer system schemes managed by governmental institutions, typically a utility. Informal settlements and the peri-urban interface, however, often rely on on-site sanitation systems, from basic pit latrines to flush toilets with septic tanks, and sometimes on water kiosks or water trucks for their water supply. These so-called 'off-the-grid' solutions are often not managed by the government, but by self-organised private stakeholders, community-based organisations or NGOs.

The characteristics of low-income settlements make the provision of basic services intrinsically very difficult and, therefore, conventional service delivery approaches are often not viable [5]. For instance, slums are often located in areas with specific physical constraints, such as low-lying ground, steep slopes or densely packed housing with very poor access via narrow and irregular pathways. When settlements are informal, key sanitation stakeholders are reluctant to invest: governmental agencies, because such area is not formal and thus not recognised; landlords, because they often do not live there; and tenants, because they are afraid of gentrification and do not want to invest without having security of tenure. This usually results in the inhabitants of such neighbourhoods paying much more for water and sanitation than people served by the government, and the services that are provided often threaten public health.

Providing sanitation services to a city as a whole invariably requires a mixture of sanitation systems, which are appropriate for different parts of the city and which can be implemented at different scales [5]. It is unlikely that the same model of service delivery will be appropriate for all areas. Therefore, a citywide sanitation plan is likely to consist of several components designed to meet the specific physical, socio-economic and service conditions for different parts of the city. The city is characterised into sanitation zones or clusters based on such aspects as topography, population density, user preferences, affordability, existing systems and/or water availability, taking into account both the existing situation and expected changes due to urbanisation. This helps to determine where on-site or off-site, networked or non-networked, dry or wet systems are most appropriate in the short- and long term.

3. Sewered or non-sewered: various sanitation systems

The previous section highlights urban diversity and the related heterogeneity in terms of sanitation status. Different contexts mean different sanitation systems, therefore it is important to adopt a sanitation system approach, as illustrated by [6]. Sanitation cannot be reduced to latrines or sewers but needs to be considered as a whole value chain, consisting of a user interface (the toilet), collection/storage (e.g. pits and septic tanks), conveyance (e.g. sewers or vacuum trucks), treatment and use and/or disposal. In particular, systems can be categorised as sewered or non-sewered, as well as dry or wet (without or with water). They can also be differentiated through their scale/domain of application: on-site (for households or buildings), small-scale (for cluster of houses or neighbourhoods) or large-scale (centralised at city level). Urban centres and western cities are usually served by a single-sewer network that conveys the wastewater to a treatment plant; we will refer to this type of system in this chapter as the 'conventional system'. However, most cities around the world still rely mainly on on-site systems, where domestic wastewater accumulates in pits and tanks in a form called 'faecal sludge' or 'septage'. Now and then, this faecal sludge needs to be pumped and transported to a faecal sludge treatment plant.

It is clear that it will not be possible to connect all urban areas to a conventional system in the foreseeable future. Nor is it desirable. The main bottleneck is of course financial: the amount of money needed to provide this is immense, and most governments do not have enough funds to build such infrastructure; besides, when such an infrastructure exists, the risk of failure is high when there is no financial means and capacities to operate and maintain the system properly. The further one is from the city centre, the more expensive it also gets. The economies of scale that can be achieved at the treatment plant level are outweighed by the dis-economies of scale of the sewer network [7]. Low-density peri-urban areas especially have a very high cost per capita. Sometimes, the geography and population density of the city make a conventional system simply unrealistic, such as in Durban, South Africa, which is characterised by low density and a hilly topography. Other cities such as Dakar, Senegal, fully assume faecal sludge management (FSM) as part of its sanitation system in its own right, cohabiting with sewers. By accepting this and by investing in FSM, the utility ends up treating a lesser volume of wastewater per capita, while avoiding the massive investments required to provide a sewer connection to all. It is the role of a good master planning exercise to determine which areas are best connected to sewers and which can be best served by FSM.

Besides the crucial system choice, there is also the choice of scale. This results from the comparison of both the costs involved, on a life-cycle basis, and the associated management schemes. For both sewered and unsewered systems, there are viable alternatives at different scales, from city to neighbourhood and the individual building level. Small-scale sanitation systems or 'decentralised systems' are seen as a promising alternative for some selected urban areas. They offer the opportunity to implement new management schemes and go beyond some of the current sanitation bottlenecks.

Total sewerage coverage may also not be desirable in cities that are water scarce. In such circumstances, it does not make sense to use large amounts of drinking water simply to flush excreta and to keep the sewer system running. Besides the current threat of water scarcity to our societies, there is also the forecasted lack of nutrients for agriculture, especially for phosphorus. While phosphorus and nitrogen are now considered treatment priorities for Western governments as they may lead to eutrophication problems in natural water bodies, they are also essential for food safety. Since these nutrients are located in urine and excreta, the sound management of resources advocates for source separation to avoid diluting them in large amounts of water, which renders their recovery or treatment both cumbersome and expensive. Source separation or even on-site treatment could offer major advantages for future sustainable urban water-management systems [8].

The following sections cover the current bottlenecks for the fast increase in sanitation coverage and how innovative sanitation planning and management can contribute to the development of more sustainable sanitation systems that are contextual, integrative and inclusive.

4. Current sanitation backlog and bottlenecks

While development priorities shift from the millennium development goals (MDGs) to the sustainable development goals (SDGs), it has to be recognised that few low- and middle-income countries were even close to reaching their sanitation MDGs. In many of them, demographic growth outweighed the progress made, despite the billions of dollars poured into the sector. A lot of infrastructure is not adapted to the context and/or not sustainable, thus failing to serve the population properly in the long run. Besides, the conventional approaches fail to reach large (and often most) parts of the population. The dearth of pragmatic answers to the need for quick increases in sanitation coverage is mainly due to the lack of an enabling environment to develop solutions that move away from the conventional system. Lüthi et al. [9] structure the enabling environment in six categories: (i) government support; (ii) legal and regulatory framework; (iii) institutional arrangements; (iv) skills and capacity; (v) financial arrangements; and (vi) socio-cultural acceptance. In the past, many water and sanitation projects failed because of the lack of an integrated approach. Sanitation programmes should take these six dimensions into consideration to ensure the long-term sustainability of sanitation infrastructure and services.

The major barriers to progress in sanitation coverage lie within the institutions, policies and realities of low- and middle-income countries [2]. The public sector is often weak in terms of skills, structures, planning capacity and bureaucratic procedures, and mechanisms are not always in place to recover investment, operation or management costs, leading to the degradation of service provision or even system failure. Depending on the political structure of the city, the division of responsibilities relating to sanitation can be an institutional headache. Responsibilities for sanitation service provision are often fragmented and sometimes overlapping among different departments and ministries. This fragmentation and overall 'poor' urban governance make coordinated action difficult and can even lead to conflict between stakeholders for resources and areas of influence.

For example, some reasons why faecal sludge management systems have not been widely implemented are the financial and political complexity involved, as well as the overlapping and unclear allocation of responsibilities and a lack of incentives for efficient operation [10, 11]. This is due to the number of stakeholders who have a financial interest in the system and also to the diversity of interests of each stakeholder. Unlike other types of infrastructure (e.g. electricity) where a single utility is usually responsible for generation, delivery, operation, maintenance and billing, a faecal sludge system is more commonly a collection of stakeholders, each of whom is responsible for a different part of the treatment chain. Dysfunctional institutional frameworks result in both a lack of accountability and disagreements between stakeholders, which can even sometimes lead to sector blockage.

The situation is similar with small-scale and on-site systems: such systems often show a mismatch with many institutional conditions (regulations, professional codes or user expectations) [12]. Many factors have been put forward to explain that conventional sewerage remains the predominant paradigm for urban sanitation delivery. There is, however, little hard evidence to say whether this is mostly due to bureaucratic or technical inertia, risk aversion, corruption (and, hence, the preference for high-cost schemes with limited local accountability), political expediency (the need to be seen to be doing something), the perception that only these systems are 'modern', or simply a lack of knowledge [13]. What is clear is that whatever incentives currently exist tend to encourage local and central authorities and their advisors to stick to conventional top-down planning and conventional centralised sewerage schemes.

Moving beyond conventional approaches towards sustainable urbanisation needs to be both top-down and bottom-up. Top-down, because it is often the only way to reform institutions, laws and regulations and bottom-up, because little can be done without dynamic individuals, communities and private sector stakeholders who have the energy, vision and creativity to innovate and validate new approaches. Flexibility is needed from governments to integrate non-governmental initiatives in their planning and to allow them to be replicated and scaled up. Above all, developing the right incentives to let initiatives grow and prevent them from being stopped by the established bureaucracy is essential.

5. The new paradigm: integration of several sanitation systems under one roof

Moving beyond business as usual necessitates leadership, vision and building partnerships. The sanitation sector is, and will remain for decades to come, under the overarching responsibility of the public sector. What can be changed is the mode of operation, adapting it to multi-stakeholder settings and fostering enabling environments. The new paradigm, which is advocated in this chapter, could materialise along different axes:

1. Building interfaces: integrated and inclusive planning implies the development of interfaces between the different sanitation systems in order to create synergies and bring them under one roof. This requires innovative management and financial schemes.

2. Participation: while conventional sewer-based system are mostly planned and implemented by a top-down approach, this is not the case with faecal sludge management or neighbourhood-level systems, where the government and utility are not the only ones to run the show. Here, the role of individuals, communities and the private sector are essential for sustainability.

3. Large-scale vision: planning for sustainable sanitation alternatives means thinking at scale from the outset with an incremental approach. Many projects fail because they remain at pilot scale and cannot reach the economies of scale necessary for their survival or replication.

4. Communication: focus on state-of-the-art data collection, analysis and communication in order to allow urban leaders and sanitation stakeholders make informed decisions.

Innovative management schemes need to provide the incentives required by urban sanitation stakeholders to change their mode of operation if the conventional paradigm of top-down planning and conventional centralised sewerage is to be challenged [13].

5.1. Towards innovative management arrangements

Proper management is above all a matter of setting the right incentives and ensuring financial sustainability. In the current situation, many governmental schemes do not provide the incentives or the financial resources for sanitation systems to be run properly, and a fortiori to be extended to low-income areas. One way of responding to this mismatch in incentives is to unbundle urban sanitation vertically and horizontally [13]. The horizontal unbundling characterises the implementation of different systems for different urban areas, as described earlier. This is a departure from one-size-fits-all solutions and allows for the use of least-cost solutions because it enables the deployment of both on-site and off-site systems in areas, depending on housing density, ground conditions, tenure and a range of other factors. Vertical unbundling recognises that communities or private sector stakeholders may be willing and able to take responsibility for part of the sanitation supply chain. Both vertical and horizontal unbundlings tend to promote decentralisation.

The result of conceptually dividing a city into management units depending on both incentives and technical feasibility is that many elements of the system can then be developed independently [13]. Community initiatives can become less dependent on city-wide actions, and financing for small elements of the system may be easier to mobilise. These reconfigurations, therefore, allow for incremental development. Notwithstanding, whatever level of delegated or decentralised management is chosen, overall government supervision is needed with proper management interfaces.

5.1.1. Management interfaces

While vertical and horizontal unbundling offer increased flexibility, they also require skilled coordination [13]. Communities or informal private stakeholders can rarely manage the whole sanitation supply chain. For example, communities may be willing to invest in a sewer network, but rarely in a treatment plant; they also often do not have the skills or the willingness to commit sufficient financial resources to build a state-of-the-art sewer network. This may result in dysfunctional networks, which in the end cannot be connected to a small-scale treatment plant, or to the city sewer system.

The creation of well-designed management interfaces is, therefore, key for sustainable urbanisation. The scaling up of small-scale wastewater treatment systems necessitates good coordination at the state/city level. Indonesia integrated community-based decentralised wastewater systems into its sanitation policy, which resulted in the implementation of thousands of systems. However, it turned out that communities alone cannot manage the treatment plants in the long run. There is a necessity for a higher structure, a 'centralised management of decentralised schemes' to carry out the minimum operation and maintenance required, to monitor the effluent quality and to take care of major troubleshooting. Scaling up entails more than replicating a large number of discrete projects [14].

Coordination is also necessary for faecal sludge management. In most cities in low- and middle-countries, there is a thriving private sector around faecal sludge/septage emptying. Unfortunately, without coordination at the city level and faecal sludge treatment infrastructure, most of the sludge ends up being discharged directly into the environment.

In Dakar, Senegal, the utility (ONAS) built faecal sludge treatment plants and integrated the faecal sludge emptying sector (privately owned vacuum trucks) into its scheme. It turned out to be a win-win situation, as the private service providers desperately needed locations to safely discharge sludge within the city. ONAS further improved the quality of the service with the creation of a call centre. This optimised the distances that the private service providers have to drive and, as a consequence, decreased the emptying fee, thus making it more affordable—a sort of 'Uber' for sludge collection.

Small-scale sanitation systems, unfortunately, still lack good examples of centralised management in low- and middle-income countries. A clear competitive advantage of most small-scale sanitation systems is that they do not require permanent staff on-site, except, perhaps, for a community member to run the pumps. Therefore, a centralised management unit at the city level, embedded within the utility, would consist of a few skilled staff, specially trained to operate small-scale systems. Such a unit would be in charge of monthly monitoring and troubleshooting, whereas the sewer network and routine maintenance (such as cleaning) would be delegated to community or neighbourhood members. Both could be connected via a call centre in case of problems. Recent technological advances in fields, such as mobile communication, chemical sensors and remote control, open up a broad and promising range of new system configurations [15].

5.1.2. Delegated management

Servicing low income and informal settlements remains a big challenge. Delegating service provision to local operators, including private companies, NGOs, CBOs, User Associations or Water Trusts is one solution for the utility to help service these areas [11, 16]. It is an effective approach, but a number of barriers must be overcome, especially technical and financial capacities and access to credit. Above all, delegated management models need to be based on clearly defined contracts which ensure benefits to all parties (consumer, local operator, utility) [16]. The utility takes on the role of a controller, setting standards and monitoring the service delivery.

Case study: Delegated management in Lusaka, Zambia

In Lusaka, Zambia, the commercial utility, the Lusaka Water and Sewerage Company (LSWC), has the mandate for water supply and sewerage delivery for the whole city. Since LWSC, however, was not serving informal peri-urban areas, two community-based Water Trusts were established by the NGO CARE to fill the gap in service delivery in Kanyama and Chazanga, two of the biggest peri-urban areas of the city. LWSC decided to integrate this initiative in its water supply service delivery scheme and formally delegated the management of the water supply in these neighbourhoods to the Water Trusts through the provision of a license. LWSC later also recognised that the conventional sewer system approach they followed in the city centre would not work in the peri-urban areas where the Trusts are working and decided to also delegate sanitation service provision to the latter. A latrine emptying service was established for the two neighbourhoods by formalising the status of informal manual emptiers already working there. These workers collect faecal sludge from pit latrines and bring it to a treatment plant. The long-term goal was for them to work as proper private entrepreneurs [17].

The Water Trusts are successfully delivering water and sanitation services due to several reasons: (i) services are adapted to the local conditions; (ii) they are well positioned and known within the community, have local staff and offices, and are therefore easily accessible; and (iii) they can support the latrine emptiers with their management capacities [18].

In the meanwhile, LWSC further recognised that the different conditions in low-income peri-urban neighbourhoods require different management schemes within their organisation and, therefore, established a department for peri-urban affairs, consisting mostly of social workers. This department is taking the lead in the coordination of these innovative delegated management schemes.

Following the example of several European countries such as Germany or France, which established decentralised management schemes in rural communities, collaborative arrangements should be sought for. A management interface also means the creation of operator networks, joint capacity building and cooperation between communities and neighbourhoods. This results in an increase in capacities and efficiency, as well as a higher level of professionalism.

5.1.3. Inclusive stakeholder involvement

Innovative management arrangements imply the involvement of stakeholders beyond the utility in the planning, implementation and operation of sanitation systems. In faecal sludge management for example, the involvement of the private—and sometimes informal—service providers is crucial; for small-scale sanitation systems at neighbourhood level, the involvement of the community is not less important. Engagement with different stakeholder groups is a critical activity that is essential for the successful development of sustainable sanitation services and behaviour change [5, 9, 19]. Enabling the civil sector and the local private sector to take a more proactive role in the definition, selection, planning and eventually management of appropriate and locally contextualised services is seen as the way forward for sustainable urban sanitation.

Stakeholder involvement in the water and sanitation sector is justified by four main arguments: ownership, efficiency, better design and empowerment. Stakeholder involvement is the art of including stakeholders in the urban planning process in order to take into account their needs, priorities and interests, to achieve consensus and to remove opposition; in other words, to make them participate. It is largely about defining the participation level of people in the process, from simple information to consultation, collaboration or delegation, and how to best answer their needs, for example through awareness raising or training and capacity building.

The benefits of alternative sanitation systems may not be clear to everybody from the beginning and some people may be reluctant to change their daily routine. For these reasons, information and transparency are fundamental. Involvement is also about showing the benefits of change to the different stakeholders and giving incentives. For example, with proper FSM, authorities gain recognition by improving the population's welfare. Informal service providers may get a voice, a status and get out of the margins of society, while the service they provide gets widely recognised. Private collection and transport entrepreneurs gain formal disposal sites and the price of services may be reduced for the households [20].

The capacity to provide services effectively and efficiently is the backbone of sustainable service provision. This includes well-trained engineers and planners at all levels (municipal, provincial and central government), but also private sector and NGO stakeholders who have their role to play. That is why capacity building and on-the-job training are crucial to improve service delivery and expand coverage in rapidly urbanising areas. The necessary capacity will need to be developed at both, the individual and collective level; individual capacity refers to particular skills that individual people have and collective capacity refers to a community or a group's capacity to organise, mobilise and support collective actions [21]. Important components of water and sanitation sector capacity building involve (i) strengthening and improving management in terms of building technical, financial and managerial capabilities; (ii) upgrading institutional and technical capacities of the key actors to help identify, understand and evaluate complex urban environmental problems; (iii) establishing co-operative partnerships with government, elected and official, civil society organisations, and the private sector to deal with cross-cutting challenges; (iv) utilising participatory tools in planning, decision-making, and political processes which facilitate the development of a common vision, articulation of needs and joint action [22].

5.2. Thinking at scale

To solve the sanitation issue in an urbanising world, decision-makers need to think at scale. Isolated initiatives carried out by dynamic entrepreneurs and civic champions are not the answer to immense sector challenges. 'Pilots never fail, but pilots never scale' (Gebauer, personal communication). Two main reasons can be mentioned for that [23]: (i) there is a tendency to overinvest in pilots in order to ensure their success, which per se makes them non-replicable; and (ii) pilots cannot reach the economies of scale which would make them competitive and sustainable, both in terms of implementation and management.

Thinking at scale is necessary to establish crucial elements, such as centralised management units or call centres. The gains driven by city-level management of FSM can be reinforced by the optimal localisation of the faecal sludge treatment plant(s), minimising the distances travelled from each neighbourhood. The centralised management of decentralised systems can only be achieved when reaching numbers from the beginning, that is starting with a critical mass of projects.

Thinking at scale allows for the development of more sustainable management schemes, which also provide incentives for the private sector. It permits the standardisation of sanitation systems, as well as the development by the government of such mechanisms as licenses and certifications, and it helps to attract investors. Indeed, many promising small-scale initiatives are not replicated because the capital needed is too low to interest the main urban sanitation donors, who are used to multi-million dollar programmes. In general, access to small amounts of credit is often a major bottleneck to sanitation stakeholders [13] and micro-entrepreneurs. In Dakar, for example, a special credit line was opened to allow faecal sludge service providers to borrow the amounts needed to renew their truck fleets.

5.2.1. Towards sustainable business models for innovative sanitation services

Some sanitation systems can constitutes an innovation in a specific context. As such, in order to be able to reach scale, new markets need to be created, highlighting the potential for the private sector and job creation. As pointed out by Truffer et al. [12], on-site treatment systems represent a major challenge to the current competencies of utilities in terms of their providing urban water management services and organising their value chain. A future large-scale application, thus, depends on the successful organisation of innovation processes in three domains [12]: (i) technological components and system integration, (ii) value chain formation and the development of new business models, and (iii) institutional innovations to create appropriate conditions under which these systems can reliably operate.

In order to reach the base of the pyramid (BoP), that is the low-income customers, innovative business models must be defined. Key factors for success are affordability, accessibility, acceptance and awareness [24]. Unfairly, low-income inhabitants in situations without proper service often have to pay more for water and sanitation than people connected to the governmental sewer system. Reasons for this are the lack of accessibility to services, transport costs, lack of economies of scale and the fact that sometimes illicit operators take advantage of the absence of viable public services. It, however, shows that there is a capacity-to-pay even in low-income neighbourhoods.

In such neighbourhoods, the sanitation challenge often starts with the lack of toilets at the household level. People have to resort to poorly maintained public toilets, shared toilets or in the worst case to open defecation. Sanitation improvements cannot be reached with the construction of new toilets only; what is needed is the development of an integrated service chain that can maintain the toilets and collect, transport and treat the excreta. In Nairobi, Kenya, the NGO Sanergy developed a system based on the use of public toilets where urine and faeces are collected in separate containers and transported to a treatment plant where they are turned into fertilisers. The service is based on a franchising system in which micro-entrepreneurs (the franchisees) decide to maintain a toilet, versus a small fee from the users. Therefore, it includes local residents and offers job creation on-site.

In Manila, Philippines, the Manila Water Company managed to incrementally extend the coverage of desludging at the household level by restructuring the tariff system. Desludging costs for households was fixed to the water volume used and not per trip as is usually the case. The tariff system created access for poor people to the desludging services, as they use less water volume. In Bangalore, this increase was triggered by the private sector itself. These businesses emerged when the Indian government was sponsoring the setting-up of toilets with pit latrines. What the Indian government did not regulate was the emptying of the pits. Some smart entrepreneurs recognised this lack of regulation and started the so-called 'honey-sucker' desludging micro-businesses (one entrepreneur, one driver and one helper). Interestingly, the micro-businesses themselves did not grow, but were replicated to more than 300 businesses. Through collective action, the micro-honey-sucker businesses developed a specific vacuum truck, which drove cost reduction from 10 to 1, customised a specific pump, and developed pricing mechanisms for apartment buildings [25, 26]. They also developed agreements with farmers to get the sludge composted and reused. In short, through their entrepreneurial approach, they created their own market and business model.

5.2.2. Centralised vs. decentralised, or how to tackle the uncertainty of urban growth

Some of the most significant advantages of decentralised sanitation systems are their flexibility, modularity and cost-effectiveness [27, 28]. They can be implemented in stages and built as close as possible to the actual wastewater volume, reducing the possibility of accruing idle capacity costs [29].

The high uncertainty in city developments and population growth in low- and middle-income countries can be addressed through a modular and incremental approach. Instead of investing large sums for treatment plant designed for a planning horizon of 30 years, several smaller plants can be built with a planning horizon of 15 years, thus serving more people in the short-term. Such an approach allows for adaptation to better meet the rise in demand, and for the avoidance of costly over-capacities. The optimal level of decentralisation must be carefully analysed for a cost-effective clustering of the city. Overall, uncertain urban growth advocates for further decentralisation of sanitation services in the decades to come.

As mentioned earlier, both vertical and horizontal unbundling tend to promote decentralisation. If wastewater treatment plants usually offer economies of scale (the bigger the plant, the lower the price per capita), it is not the case for sewer networks, which typically represent more than 80% of the investment costs of a sanitation system [30], and also a significant part of the operation costs. The predominant 'expand and upgrade' leads to biased economic incentives because stakeholders tend to base their decisions on economies of scale in the cost of a centralised wastewater plant, while neglecting economies of scale at the level of the entire network, which are, as a rule, much more difficult to assess [31]. The optimum configuration will generally be defined by some sort of hybrid constellation [32], also referred to as a distributed wastewater infrastructure [33]. The strong reliance of most utility services on centralised network infrastructure is also becoming increasingly challenged by new technological advances in decentralised alternatives, as well as remote operations [28]. Eggimann et al. [34] confirmed that the optimal degree of centralisation decreases with increasing terrain complexity and settlement dispersion, while showing that the effect of the latter exceeds that of topography. The use of more decentralised sewer systems or even simplified sewer systems can allow for major savings, by reducing the number of pumps and force mains, the depth of sewers and the size of manholes and the gradient of the pipes [6]. Overall, it also drastically reduces the operation costs. In case of organised communities, such sewer systems can be partly built and managed by the community itself.

5.3. Communication: understanding and visualising the situation

A clear overview of the diverse sanitation challenges in a city is the basis for the development of sound urban sanitation strategies. This requires good quality data and awareness from the decision makers. Accessing relevant data is challenging, especially in contexts where data is scarce and the urban development very dynamic. Often, data are either not collected or analysed properly, or, sometimes, hidden or manipulated for political or personal reasons. Governmental agencies usually have some reports, statistics and maps that can serve as a preliminary introduction. However, they should always be considered with care and, therefore, the collection of primary data is recommended. In unplanned and informal neighbourhoods without legal status, there is often no official data and the collection of primary data can be the only way of assessing the urban environment. The best way to get a reasonably accurate estimation is to rely on several sources of information, which can be cross-checked and, if needed, complemented by further research [35].

All relevant urban sanitation stakeholders should be consulted in the planning process, such as water and sewerage utilities, the private service providers and the end-users. This helps the understanding of the heterogeneous urban environment and gives access to first-hand information from different perspectives. Depending on the stakeholder, data can be collected through different tools, such as household surveys and expert interviews, as well as other participatory methods that are focus group discussions, town hall meetings, transect walks or participatory mapping [5, 9].

New technologies facilitate data collection and visualisation. Mobile data collection and geo-referencing tools quicken the process, increase the quality of the data collected and allow rapid visualisation of complex urban areas. Mobile data collection is gaining importance as data can be collected with any mobile phone and viewed in real time [36]. With the decreasing costs and complexity of geospatial data collected from satellites, these data become accessible not only to all urban planners but also to civil society, which can even participate in the monitoring efforts. This is often already done, for instance, with the monitoring of water points.

An innovative way to visualise outcomes and communicate the urban sanitation challenge at the city-wide level are the so-called 'shit/excreta flow diagrams' (SFDs), which clearly show how excreta is or is not contained as it moves along multiple pathways from defecation to disposal or end-user. As illustrated in **Figure 2** for the city of Dar es Salaam, Tanzania, the SFD

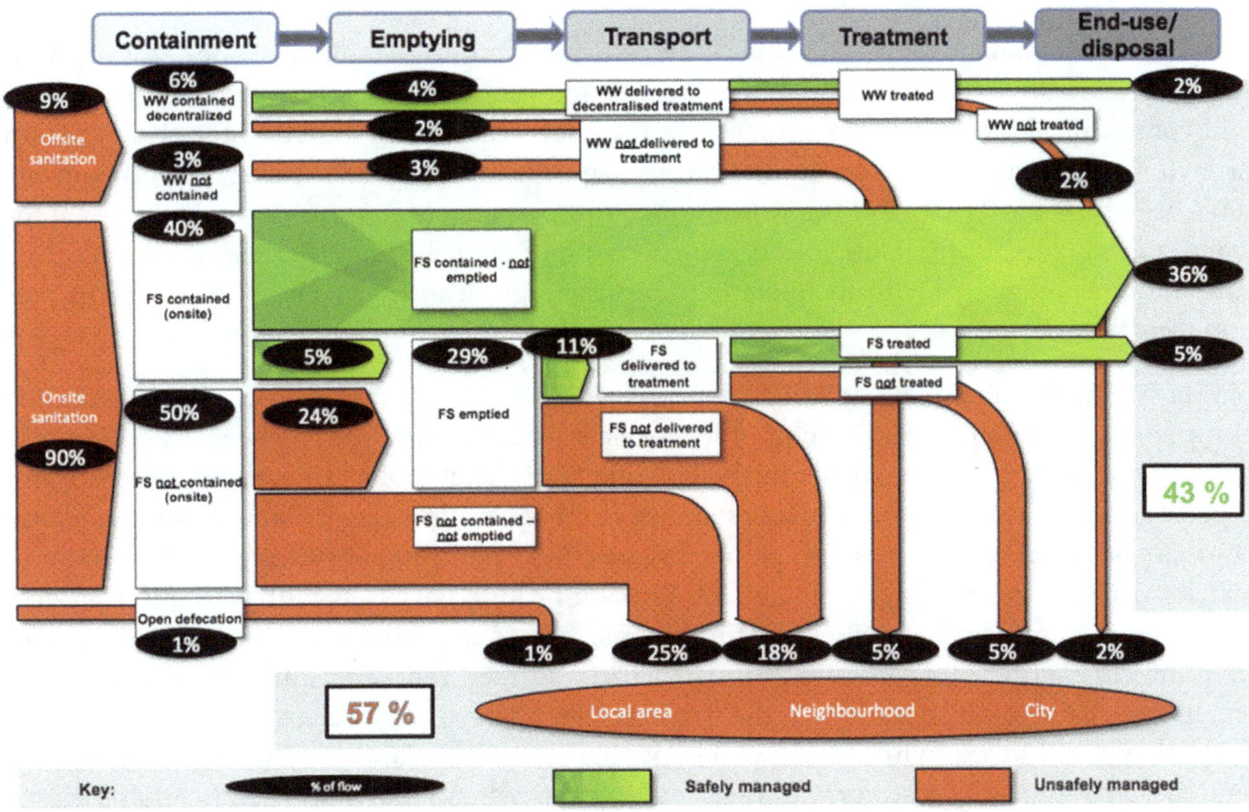

Figure 2. The excreta flow diagram for Dar Es Salaam, Tanzania, shows the proportion of faecal matter that is safely managed (in green) compared to unsafely managed streams (in red) [37].

is an advocacy and decision-support tool that can easily be understood by non-technical key stakeholders and by civil society. It, therefore, has the potential to shift the focus of attention, money and activities towards more effective and inclusive urban sanitation beyond water-borne sewerage.

6. Sanitation in the urban ecosystem

Integrating sanitation in the urban ecosystem, that is seeing it as more than merely collecting and treating wastewater, is seen as a main driver or incentive for more efficient and sustainable sanitation services. This can take different forms: (i) synergies between different services (e.g. energy, communication); (ii) multi-functional sanitation concepts; (iii) urban valorisation of sanitation end products (e.g. treated wastewater, nutrients, heat or biogas).

Good city sanitation plans recognise the links between sanitation and other municipal services [5]. For example, uncollected solid waste ends up in drains and sewers, greatly increasing maintenance requirements. Consideration of the integration between these different services is important to ensure effective sanitation service delivery. Storm water drainage also needs to be planned in parallel, as neglecting it can quickly lead to the collapse of the sanitation system, especially combined with the lack of solid waste management.

A utility which manages different urban services may be more sustainable if one of these public services is lucrative. For example, in Cuenca, Ecuador, the utility manages water, sanitation and communication. In that case, the communication sector contributed to the development of inclusive and state-of-the art sanitation services, through cross-financing. This would not have been possible if both services were managed separately. Managing several services at the same time can also support the often delicate issue of fee collection. For example, collecting sanitation fees together with the electricity bill can significantly increase the collection rate, and thus the cost recovery.

Similarly, multi-functional sanitation concepts increase the sustainability of the sanitation component through the provision of other services. Multi-functional public toilet concepts in low-income settlements have made their proof, like the Ikotoilet in Kenya, which brings together toilets, showers, sale of personal care products and, in collaboration with local companies, supports advertising, telephony, and shoe-polishing services as well as sales of snacks, drinks and newspapers. In general, linking productive assets with sanitation services increases sustainability. Activities needing biogas, such as communal kitchens, can for instance be built in synergy with sanitation infrastructure in green markets; similarly, decentralised treatment plants can be built so that the treated effluent can directly be used by urban farmers for irrigation. Linked with solid waste management and landscaping, multi-functional sanitation concepts can also contribute to increase the recreational value of a neighbourhood.

The valorisation of sanitation end products should be fostered. In high-income countries, grey water recycling in buildings, on-site reuse of treated wastewater for urban gardening and heat recovery from sewage are increasingly being implemented, as well as the biogas production in centralised treatment plants for electricity production or as a fuel for public transportation. Technological advances, increasing scarcity of water and rising prices of fuel and fertilisers will soon make such techniques more affordable and competitive in low- and middle-income areas as well.

7. Conclusion

Urban sanitation is at an inflection point: although conventional sewer systems and large-scale treatment plants are still seen as the golden standard, there is an increasing acknowledgement that sanitation systems such as faecal sludge management or decentralised/small-scale systems are complementary options and constitute viable alternatives in the long run for selected parts of cities. The change in thinking, however, still has to reach all the city sanitation stakeholders to become a widespread and operating reality, from the urban leaders to the service providers and the engineers and consultants. The multi-disciplinary nature of the new approaches makes them more complex than the engineering-centric one. In order to reach full sanitation coverage in low-and middle-income countries, the sector should be enriched with the social and economic skills that are needed to involve the different sanitation stakeholders in planning processes and devise innovative management and financial schemes.

Although going beyond the current bottlenecks may seem daunting, it is possible to change the situation by creating the right incentives for the different stakeholders. The new paradigm advocated for in this chapter is in line with that: building interfaces, delegating management through arrangements which ensure benefits for all parties, increasing participation, thinking at scale and fostering better communication all contribute to reducing the sanitation burden of the governmental institutions and improving decision making, while increasing sanitation coverage and thus the consumer basis. There are ways to structure the sector through better governance, appropriate planning tools, the involvement of private stakeholders and innovative financing mechanisms.

Sanitation has a key role to play in sustainable urbanisation, and it needs to be fully integrated in the current thinking about 'circular,' 'efficient,' 'green' or 'water-secure' cities. Sanitation, as the supply chain for both used water and nutrients, is at the core of these urban sustainability concepts. Cities of the future will probably treat wastewater and sludge as close as possible to their source, to enable on-site reuse of resources. Many pilot projects already go in this direction, by fostering grey water recycling in buildings and use of treated wastewater in urban gardening. In water-scarce regions, there is a growing concern to plan wastewater treatment infrastructure so that it can directly contribute to groundwater recharge and safe water supply.

Integrating several sanitation systems under one roof, as has been outlined in this chapter is a first necessary step for successful sanitation coverage in rapidly urbanising cities. The better integration of sanitation in the 'urban ecosystem' is the promising second step, contributing to two of the main drivers of change: reducing the costs through better synergies among services, while providing more incentives to the sanitation stakeholders.

Author details

Philippe Reymond*, Samuel Renggli and Christoph Lüthi

*Address all correspondence to: philippe.reymond@eawag.ch

Eawag: Swiss Federal Institute of Aquatic Science and Technology, Duebendorf, Switzerland

References

[1] United Nations, Department of Economic and Social Affairs, Population Division (UNDESA). The World Population Prospects: 2015 Revision [Internet]. 2015. Available from:http://www.un.org/en/development/desa/publications/world-population-prospects-2015-revision.html [Accessed: 01.04.2016].

[2] Mara D, Alabaster G. A new paradigm for low-cost urban water supplies and sanitation in developing countries. Water Policy. 2008;10(2):119–129. doi:10.2166/wp.2008.034.

[3] Foster V, Briceno C. Africa's infrastructure: a time for transformation. Washington DC, US: World Bank; 2010. doi:10.1596/978-0-8213-8041-3.

[4] Lüthi C, Panesar A, Schütze T, editors. Sustainable sanitation in vities – a framework for action. 1st ed. The Netherlands: Papyroz Publishing House; 2011. 148p. doi: 10.1.1.365.6104.

[5] Parkinson J, Lüthi C, Walther D. Sanitation 21 – a planning framework for improving city-wide sanitation services. 1st ed. London, UK: IWA; 2014. 38p.

[6] Tilley E, Ulrich L, Lüthi C, Reymond P, Zurbrügg C. Compendium of sanitation systems and technologies. 2nd ed. Dübendorf, Switzerland: Eawag; 2014. 176p.

[7] Pinkham R D, Hurley E, Lovins A, Magliaro J, Watkins K. Valuing small-scale wastewater technologies: a catalog of benefits, costs, and economic analysis techniques. 1st ed. Colorado, US: Rocky Mountain Institute; 2004.Available from: http://www.rmi.org/Knowledge-Center/Library/W04-21_ValuingDecentralizedWastewater [Accessed: 18.05.2016]

[8] Larsen T A, Gujer W. Implementation of source separation and decentralization in cities. In: Larsen T A, Udert K M, Lienert J, editors. Source separation and decentralization for wastewater management. 1st ed. London, UK: IWA Publishing; 2013. p. 135–149.

[9] Lüthi C, Morel A, Tilley E, Ulrich L. Community-led urban environmental sanitation, complete guidelines for decision-makers with 30 tools. 1st ed. Dübendorf, Switzerland: Eawag; 2011. 100p.

[10] Tilley E, Dodane P. Financial transfers and responsibility in faecal sludge management chains. In: Strande L, Ronteltap M, Brdjanovic D, editors.Faecal Sludge Management. Systems Approach for Implementation and Operation. 1st ed. London, UK: IWA Publishing; 2014. p. 273–291.

[11] Bassan M. Institutional frameworks for faecal sludge management. In: Strande L, Ronteltap M, Brdjanovic D, editors. Faecal sludge management. Systems approach for implementation and operation. 1st ed. London, UK: IWA Publishing; 2014. p. 255–271.

[12] Truffer B, Binz C, Gebauer H, Störmer E. Market success of on-site treatment: a systemic innovation problem. In: Larsen T A, Udert K M, Lienert J, editors. Source separation and decentralization for wastewater management. 1st ed. London, UK: IWA Publishing; 2013. p. 209–223.

[13] Evans B. Sanitation in cities of the global south – is decentralisation a solution? In:Larsen T A, Udert K M, Lienert J, editors. Source separation and decentralization for wastewater management. 1st ed. London, UK: IWA Publishing; 2013. p. 117–131.

[14] Eales K, Siregar R, Febriani E, Blackett I. Review of community managed decentralized wastewater treatment systems in Indonesia. 1st ed. Indonesia: WSP; 2013.

[15] Olsson G. The potential of control and monitoring. In: Larsen T A, Udert K M, Lienert J, editors. Source separation and decentralization for wastewater management. 1st ed. London, UK: IWA Publishing; 2013. p. 179–191.

[16] Water & Sanitation for the Urban Poor (WSUP). The urban programming guide – how to design and implement an effective urban WASH programme. 1st ed. London, UK: WSUP; 2014. 53p.

[17] Mulenga M. Faecal sludge management by a water trust in Zambia – the case of Kanyama, Lusaka. In: FSM3; 18–22 January 2015; Hanoi, Vietnam. IWA; 2015.

[18] Water & Sanitation for the Urban Poor (WSUP). FSM services in Lusaka: moving up the excreta management ladder [Internet]. 2014 [Updated: 2014]. Available from: http://www.wsup.com/resource/fsm-services-in-lusaka-moving-up-the-excreta-management-ladder/ [Accessed: 06.04.2016].

[19] Reymond P. Planning integrated faecal sludge management systems. In: Strande L, Ronteltap M, Brdjanovic D, editors. Faecal sludge management. Systems approach for implementation and operation. 1st ed. London, UK: IWA Publising; 2014. p. 363–387.

[20] Reymond P, Bassan M. Stakeholder engagement. In: Strande L, Ronteltap M, Brdjanovic D, editors. Faecal sludge management. Systems approach for implementation and operation. 1st ed. London, UK: IWA Publishing; 2014. p. 341–362.

[21] Hamdi N, Goethert R. Action planning for cities. Community practice. 1st ed. Chichester, UK: Wiley; 1997.

[22] Markard J, Parkinson J. Putting Plans into Practice. In: Lüthi C, Panesar A, Schütze T, editors. Sustainable sanitation in cities – a framework for action. 1st ed. The Netherlands: Papyroz Publishing House; 2011. p. 148.

[23] Reymond P, Abdel Wahaab R, Moussa M. Policy recommendations for the scaling-up of small-scale sanitation in Egypt. 1st ed. Cairo, Egypt: Eawag; 2015. 40p.

[24] Anderson J, Markides C. Strategic innovation at the base of the economic pyramid. 1st ed. Germany, UK: European School of Management and Technology, Germany; London Business School, UK; 2006.

[25] Kvarnström E, Verhagen J, Nilsson M, Srikantaiah V, Ramachandran S, Singh K. The business of the honey-suckers in Bengaluru (India): the potentials and limitations of commercial faecal sludge recycling – an explorative study. 1st ed. The Hague: IRC International Water and Sanitation Centre; 2012.

[26] Hystra. Designing the next generation of sanitation businesses [Internet]. 2014 [Updated: 2014]. Available from: http://hystra.com/sanitation/ [Accessed: 06.04.2016].

[27] Massoud M A, Tarhini A, Nasr J A. Decentralized approaches to wastewater treatment and management: applicability in developing countries. Journal of Environmental Management. 2009;90(1):652–659. doi:10.1016/j.jenvman.2008.07.001.

[28] Libralato G, Volpi Ghirardini A, Avezzù F. To centralise or to decentralise: an overview of the most recent trends in wastewater treatment management. Journal of Environmental Management. 2012;94(1):61–68. doi:10.1016/j.jenvman.2011.07.010.

[29] Maurer, M. Specific net present value: an improved method for assessing modularisation costs in water services with growing demand. Water Research. 2009;43(8): 2121–2130. doi:10.1016/j.watres.2009.02.008.

[30] Maurer M, Rothenberger D, Larsen T A. Decentralised wastewater treatment technologies from a national perspective: at what cost are they competitive? Water Science and Technology: Water Supply. 2005;5(6):145–154.

[31] Maurer M, Scheidegger A, Herlyn A. Quantifying costs and lengths of urban drainage systems with a simple static sewer infrastructure model. Urban Water Journal. 2013;10(4):268–280. doi:10.1080/1573062X.2012.731072.

[32] Poustie M S, Deletic A, Brown R R, Wong T, de Haan J, Skinner R. Sustainable urban water futures in developing countries: the centralised, decentralised or hybrid dilemma. Urban Water Journal. 2015;12(7):543–558. doi:10.1080/1573062X.2014.916725.

[33] Tchobanoglous G, Leverenz H. The rationale for decentralization of wastewater infrastructure. In: Larsen T A, Udert K M, Lienert J, editors. Source separation and decentralization for wastewater management. 1st ed. London, UK: IWA Publishing; 2013. p. 101–115.

[34] Eggimann S, Truffer B, Maurer M. To connect or not to connect? Modelling the optimal degree of centralisation for wastewater infrastructures. Water Research. 2015;84(1): 218–231. doi:10.1016/j.watres.2015.07.004.

[35] Reymond P. Assessment of the initial situation. In: Strande L, Ronteltap M, Brdjanovic D, editors. Faecal sludge management. Systems approach for implementation and operation. 1st ed. London, UK: IWA Publishing; 2014. p. 295–318.

[36] Hutchings M, Dev A, Palaniappan M, Srinivasan V, Ramanathan N, Taylor J. mWASH: mobile phone applications for the water, sanitation, and hygiene sector. 1st ed. Oakland, US: Pacific Institute; 2012. 115p.

[37] Eawag. Improving understanding of urban sanitation [Internet]. 2016 [Updated: 2016]. Available from: http://www.sfd.susana.org/ [Accessed: 06.04.2016].

Permissions

All chapters in this book were first published in SU, by InTech Open; hereby published with permission under the Creative Commons Attribution License or equivalent. Every chapter published in this book has been scrutinized by our experts. Their significance has been extensively debated. The topics covered herein carry significant findings which will fuel the growth of the discipline. They may even be implemented as practical applications or may be referred to as a beginning point for another development.

The contributors of this book come from diverse backgrounds, making this book a truly international effort. This book will bring forth new frontiers with its revolutionizing research information and detailed analysis of the nascent developments around the world.

We would like to thank all the contributing authors for lending their expertise to make the book truly unique. They have played a crucial role in the development of this book. Without their invaluable contributions this book wouldn't have been possible. They have made vital efforts to compile up to date information on the varied aspects of this subject to make this book a valuable addition to the collection of many professionals and students.

This book was conceptualized with the vision of imparting up-to-date information and advanced data in this field. To ensure the same, a matchless editorial board was set up. Every individual on the board went through rigorous rounds of assessment to prove their worth. After which they invested a large part of their time researching and compiling the most relevant data for our readers.

The editorial board has been involved in producing this book since its inception. They have spent rigorous hours researching and exploring the diverse topics which have resulted in the successful publishing of this book. They have passed on their knowledge of decades through this book. To expedite this challenging task, the publisher supported the team at every step. A small team of assistant editors was also appointed to further simplify the editing procedure and attain best results for the readers.

Apart from the editorial board, the designing team has also invested a significant amount of their time in understanding the subject and creating the most relevant covers. They scrutinized every image to scout for the most suitable representation of the subject and create an appropriate cover for the book.

The publishing team has been an ardent support to the editorial, designing and production team. Their endless efforts to recruit the best for this project, has resulted in the accomplishment of this book. They are a veteran in the field of academics and their pool of knowledge is as vast as their experience in printing. Their expertise and guidance has proved useful at every step. Their uncompromising quality standards have made this book an exceptional effort. Their encouragement from time to time has been an inspiration for everyone.

The publisher and the editorial board hope that this book will prove to be a valuable piece of knowledge for researchers, students, practitioners and scholars across the globe.

List of Contributors

Gökçen Kılınç Ürkmez
Bursa Technical University, City and Regional Planning Department, Bursa, Turkey

Claudia Trillo
School of the Built Environment, UPRISE University of Salford, The Crescent, Salford, Greater Manchester, UK

Onur Şatir
Department of Landscape Architecture, Faculty of Agriculture, Yuzuncu Yil University, Van, Turkey

Selma Çelikyay
Department of Landscape Architecture, Faculty of Forestry, University of Bartın, Bartın, Turkey

Okan Murat Dede
Amasya University, Amasya, Turkey

Ilaria Giovagnorio and Giovanni M. Chiri
DICAAR, University of Cagliari, Cagliari, Sardinia, Italy

Ebru Ersoy
Department of Landscape Architecture, Faculty of Agriculture, Adnan Menderes University, Aydin, Turkey

Simone F. Teixeira
University of Pernambuco, Recife, Pernambuco, Brazil
Research Group on Ethnoecology and Tropical Fish Ecology Studies of University of Pernambuco, Recife, Pernambuco, Brazil

Daniele Mariz and Anna Carla F. F. de Souza
Research Group on Ethnoecology and Tropical Fish Ecology Studies of University of Pernambuco, Recife, Pernambuco, Brazil
Federal University of Pernambuco, Recife, Pernambuco, Brazil

Susmara S. Campos
Research Group on Ethnoecology and Tropical Fish Ecology Studies of University of Pernambuco, Recife, Pernambuco, Brazil
Federal Institute of Education, Science and Technology of Pernambuco, Recife, Pernambuco, Brazil

Dong Jiang, Jingying Fu and Gang Lin
State Key Laboratory of Resources and Environmental Information System, Institute of Geographical Sciences and Natural Resources Research, Chinese Academy of Sciences, Beijing, China

Philippe Reymond, Samuel Renggli and Christoph Lüthi
Eawag: Swiss Federal Institute of Aquatic Science and Technology, Duebendorf, Switzerland

Index

T

U

V

W

Z

www.ingramcontent.com/pod-product-compliance
Lightning Source LLC
LaVergne TN
LVHW061509220926
848485LV00017B/599